SpringerBriefs in Applied Sciences and Technology

SpringerBriefs present concise summaries of cutting-edge research and practical applications across a wide spectrum of fields. Featuring compact volumes of 50 to 125 pages, the series covers a range of content from professional to academic.

Typical publications can be:

- A timely report of state-of-the art methods
- An introduction to or a manual for the application of mathematical or computer techniques
- A bridge between new research results, as published in journal articles
- A snapshot of a hot or emerging topic
- An in-depth case study
- A presentation of core concepts that students must understand in order to make independent contributions

SpringerBriefs are characterized by fast, global electronic dissemination, standard publishing contracts, standardized manuscript preparation and formatting guidelines, and expedited production schedules.

On the one hand, **SpringerBriefs in Applied Sciences and Technology** are devoted to the publication of fundamentals and applications within the different classical engineering disciplines as well as in interdisciplinary fields that recently emerged between these areas. On the other hand, as the boundary separating fundamental research and applied technology is more and more dissolving, this series is particularly open to trans-disciplinary topics between fundamental science and engineering.

Indexed by EI-Compendex, SCOPUS and Springerlink.

W. J. Nuttall · M. A. Grzegorczyk-Nuttall ·
V. W. E. Marsh · C. P. Spedding · D. Webbe-Wood ·
T. B. Scott

Fusion on the Moon

Nuclear Fusion for High-power, Semi-independent Lunar Development

Springer

W. J. Nuttall
School of Electrical, Electronic and Mechanical Engineering
University of Bristol
Bristol, UK

V. W. E. Marsh
Bristol, UK

D. Webbe-Wood
London, UK

M. A. Grzegorczyk-Nuttall
College of Business
University of Leicester
Leicester, UK

C. P. Spedding
Leverhulme Centre for Humanity and Space
University of Leicester
Leicester, UK

T. B. Scott
School of Physics
University of Bristol
Bristol, UK

ISSN 2191-530X ISSN 2191-5318 (electronic)
SpringerBriefs in Applied Sciences and Technology
ISBN 978-3-032-38216-0 ISBN 978-3-032-38217-7 (eBook)
https://doi.org/10.1007/978-3-032-38217-7

This work was supported by University of Bristol.

This Springer imprint is published by the registered company Springer Nature Switzerland AG
The registered company address is: Gewerbestrasse 11, 6330 Cham, Switzerland

Acknowledgements The authors are most grateful to the United Kingdom Atomic Energy Authority for sponsoring a related study in early 2025 and later for sponsoring this book. The ideas presented in this book, in some cases go beyond those considered in earlier work.

The authors are also most grateful to two critical readers who read a draft of the 2025 report and offered very helpful comments. One such reader was Dr. Mike Bluck of the Centre for Nuclear Engineering at Imperial College London, and the other has asked to remain anonymous, but we can report that their input concerned space issues, in particular.

The authors alone are responsible to the ideas in this book and the sponsor, and the critical readers are not responsible for the content. Ideas and conclusions presented in this book are not necessarily shared by our sponsors or advisors.

Competing Interests The authors have no competing interests to declare that are relevant to the content of this manuscript.

Disclaimers

Liability: The authors and the publisher cannot assume responsibility for the validity of all materials or for the consequences of their use. This book contains no advice or guidance and should not be used as the basis of any investment or other decision.

Trademarks: Product or corporate names may be trademarks or registered trademarks, and are used only for identification and explanation without intent to infringe.

Images: Efforts have been made to establish and contact copyright holders for all images presented in this book. We are most grateful to all the various rights holders who have kindly granted permission for reproduction. Despite our best endeavours, there may be instances where the rights of third parties have been overlooked. In such cases, we apologise and we ask that rights holders make contact and we will endeavour to resolve matters.

Contents

About the Authors

William Nuttall CPhys FInstP is Professor of Energy Policy and Transformation (part-time) at the University of Bristol, UK. He is also a non-resident fellow of the Payne Institute for Public Policy at the Colorado School of Mines, USA, and a Fellow (honorary) of Cambridge Judge Business School. He is an Affiliated Lecturer in Cambridge University Engineering Department and a Fellow of Hughes Hall, University of Cambridge. He holds a Ph.D. in Physics (1993) from the Massachusetts Institute of Technology, USA, where he was a Fulbright scholar (1987–1988).

Małgorzata Grzegorczyk-Nuttall is a lecturer in Marketing and Innovation at the University of Leicester, UK. She holds a Ph.D. in Economics from the University of Lodz, Poland, and was a Fulbright Scholar at the University of Texas at Austin. Her expertise includes technology transfer, commercialisation and research-industry collaboration. She is the author of *Knowledge and Technology Transfer from Research Organisations to Businesses*, and, *Marketing in University-Industry Collaborations,* two books that reflect her longstanding interest in how science is translated into practice. She is currently focussed on the space sector, especially future commercial scenarios for the lunar economy.

Vidar W. E. Marsh is a physics postgraduate researcher at the University of Bristol with strong computing skills and an interest in data-centric physics research. He came to research from a first-class Masters in physics from the University of Bristol. He has experience transferring his scientific and research skills to industry settings. One previous example project involved the LHCb detector alignment codebase at the large European particle physics laboratory, CERN.

C. P. Spedding is a Research Associate at the Leverhulme Centre for Humanity in Space, working at the interface of emerging technology, security, nuclear weapons, and nuclear technologies. He is an interdisciplinary researcher with master's degrees in Mechanical Engineering (MEng, Sheffield Hallam University) and Nuclear Energy (MRes, Imperial College London), and a Ph.D. in the Governance of Outer Space from The Open University. Prior to joining the LCHS, Chris was the Emerging Technologies Programme Manager, and Growth and Impact Manager at the British American Security Information Council (BASIC) and was also a civil servant for a time. His research interests include the evolving character of nuclear competition and conflict, and specifically the role of emerging and evolving technologies in that.

David Webbe-Wood received his Ph.D. from the Open University in 2024. Prior to that he worked for UK Civil Service in a number of scientific and engineering roles. He has participated in the IAEA EMRAS and BIOMOVS biosphere modelling programmes and the Anglo-Indian eMEANNS project for which he is a co-author of the book Perspectives on Engineering Uncertainty.

Tom Scott is Professor in Materials and Director of the Interface Analysis Centre at the University of Bristol, a specialist analytical centre for research on materials of all types. He holds a first-class MSci honours degree in Geology and a Ph.D. in Nuclear Materials. He is also a Royal Academy of Engineering Professorial Research Fellow co-funded by the UKAEA. He is Director of the South West Nuclear Hub and lectures in the Schools of Physics and Earth Sciences at Bristol. He has expertise in the detection and characterisation of nuclear materials to aid prediction of their behaviour in engineered and environmental scenarios. This includes activities relating to robotics and radiation detection for both fission and fusion energy applications. He works closely with UK government departments and agencies to ensure that facilities, materials and wastes stay safely managed.

Glossary

CLPS	Commercial Lunar Payload Surfaces
CTE	Coefficient of Thermal Expansion
ESA	European Space Agency
HLS	Human Landing System
HTS	High Temperature Superconductor
ISRU	In Situ Resource Utilisation
ISS	International Space Station
ITER	International Thermonuclear Experimental Reactor[1]
JET	Joint European Torus
JWST	James Webb Space Telescope
LH3M	Lunar Helium Mining LLC
MWPT	Microwave Wireless Power Transmission
NASA	National Aeronautics and Space Administration (USA)
PSR	Permanently Shadowed Regions
R&D	Research and Development
RTG	Radioisotope thermoelectric generator
ST	Spherical Tokamak
STEEPLE	Social, Technological, Economic, Environmental, Political, Legal and Ethical
UHV	Ultra-High Vacuum
UKAEA	United Kingdom Atomic Energy Authority
UN	United Nations
VC	Venture Capital

[1] Since July 2001 "Iter" is no longer an acronym. It is now officially regarded as simply a name. It is inspired by the Latin word for 'the path'.

Chapter 1
Context for Fusion on the Moon

Contents

1.1 What Is Nuclear Fusion Energy?

This book deals with the possibilities for simultaneous developments in two of the most challenging high technology areas: space exploration and nuclear fusion energy.

Nuclear fusion energy represents the culmination of a human journey spanning recorded time. It is a journey towards ever more intense sources of heat, starting from the burning of wood and later coal to the use of uranium in the late twentieth century. Fusion energy is the culmination of the journey. It uses a lightweight fuel to produce vast amounts of useful energy, more even than that available from the physics of fission that powers today's nuclear power plants.

W. J. Nuttall et al., *Fusion on the Moon*,
SpringerBriefs in Applied Sciences and Technology,
https://doi.org/10.1007/978-3-032-38217-7_1

The physics of fusion energy relates closely to the cosmos as it is nuclear fusion that powers the stars, including our own sun. While it might be poetic to report that the leading orthodox candidate energy source for lunar settlement—solar photovoltaic panels is also, in a way, a form of fusion on the Moon, that is not our focus. Rather we focus explicitly on the use of controlled nuclear fusion reactions to provide a power source capable of sustaining a semi-independent human community on the Moon. That community is likely, we suggest, to be focussed, at least initially, on industrial matters of resource extraction.

Nuclear fusion energy was first released by human beings in the Ivy Mike test of the physics of thermonuclear weapons conducted by the United States on 1 November 1952, at Enewetak Atoll in the Marshall Islands of the Pacific Ocean. That test demonstrated that human beings could recreate the energy source of the stars. In the 80 years that followed the physics has been confirmed and developed and engineering has sought to deliver the new energy source in ways that can be sustained and controlled. In the late 2020s, many private companies worldwide have joined with long-established publicly funded laboratories in seeking to deliver commercial fusion energy [1].

The Moon is an environment where much effort and cost is required to deliver engineered devices and fuels. As such, there is something intrinsically natural in seeking to explore the possible use of an energy source with more potential to do useful work per gram of fuel than any other alternative. In Chap. 3, we shall describe the engineering of the leading concept for a fusion energy device capable of producing useful energy for sustained periods. That device is known as a 'tokamak'.

1.2 The Case for Going to the Moon

While the Moon has not been visited by humans since Apollo 17 in 1972, there are several oft-cited reasons to return in the coming years. Beyond scientific curiosity, there are opportunities for exploration, economic development, and potential long-term survival strategies for humanity.

In this book, we look to the future—roughly 100 years ahead. However, our main timeframe of interest is the period 2040–2070. This is the timescale on which it will become clear whether fusion on the Moon has a long-term future. We are not saying that people must think about fusion as an option starting in 2040, but rather that if they start such thinking now, that is the earliest credible date by which tangible achievements on the Moon might be imagined.

Despite being Earth's closest neighbour, the Moon's surface has been visited only sparingly. To date, there have been 27 missions that achieved a successful soft landing on the lunar surface since this feat was first achieved by the Soviet Union's Luna 9 mission in 1966 [2]. Only six of the 27 missions carried humans. Moreover, only nine such missions have occurred in the last four decades to varying degrees of success [2]. Despite nearly 60 years of exploration, many scientific questions about the Moon remain unanswered. While one can argue that modern robotic missions can match the scientific data collection capabilities of human explorers, it is the unique human potential to see the unexpected, or to exploit serendipity, that motivates the scientific presence of humans on the Moon. Returning to the lunar surface offers an opportunity to improve our understanding of its geology and formation which in turn provides further insights into the history of the Solar System. Additionally, the far side of the Moon offers a unique opportunity for constructing large telescopes to study space weather phenomena such as solar and cosmic radiation in an environment that is not affected by the Earth's atmosphere or magnetosphere. Notwithstanding opportunities for robotic experiments in low Earth orbit, the Moon offers an opportunity to study low gravity influences on biological processes and materials science, especially after the International Space Station is decommissioned in 2030. The ability to operate at scale and for human scientists to interact directly with the experiments is potentially valuable.

Beyond scientific research, the Moon offers a unique testing ground for deep space missions. Its proximity to Earth makes it an ideal location for trialling new technologies, life support systems, and operational procedures, before using them in more ambitious human exploration missions to Mars and beyond. Unlike these far destinations which take months, or years, to reach, the Moon's proximity to the Earth means that a rescue mission could arrive within days in an emergency, with the right preparation. This majorly de-risks human space exploration and helps make crewed missions to Mars more attractive when compared to a direct-to-Mars mission. A key difference between the Moon and Mars is that Mars has a low-pressure atmosphere. That atmosphere is sufficient to generate dust storms and other local weather. In a lunar context, the vast majority of dust-generating disruption will come from human activity and hence is within human control. There is no risk of weather on the Moon. With that said however, there are numerous risks on the Moon that we are protected from on Earth by our atmosphere and the Earth's geomagnetic core. Such risks will be returned to in Sect. 2.3.

The Moon is also interesting for strategic reasons, both as a spaceport for deep space exploration and as an insurance policy for the survival of humanity. It has been confirmed that the Moon contains water-ice hidden in the permanently shadowed regions (PSRs) in, for example, lunar craters [3].

The idea of water on the Moon was brought to popular attention by the Belgian author and illustrator Hergé in his Tintin adventure *Explorers on the Moon* published in the early 1950s. As part of his extensive research for his pre-Sputnik tale, Hergé discussed the technical aspects with many experts including Max Hoyaux a scientist at Belgium's ACEC nuclear research centre [4]. Hoyaux alerted Hergé to the possibility of water on the Moon. Hergé developed the idea and he has Tintin and colleagues exploring a lunar cave, before Tintin falls gently onto ice (see Fig. 1.1). The American version of the comic has Tintin cry out "Yikes!... Ice!" while the British text has him exclaiming "Great Snakes! ... Ice!". In the 1950s the bold idea of ice on the Moon was met with some initial scepticism, but that scepticism has long since evaporated.

Returning to the twenty-first century, lunar water could be used to synthesise rocket fuel and support a permanent lunar settlement. This water would allow the

Fig. 1.1 Tintin falls onto ice in a lunar cave. Hergé, Explorers on the Moon (page 36, vignette C6). Image rights: © Hergé-Tintinimaginatio 2026 with permission

Moon to function as a space port for deeper space exploration which has the following benefits: launching from the Moon requires significantly less energy to break free from the Earth's gravitational influence compared to launch from the Earth. This means that rockets need to carry significantly less fuel to reach their destinations, enabling deeper space missions with a large payload capacity. Counter to this, we must concede that there is a case to be made for a free-standing space port to be constructed at a Lagrange Point where the gravity well would be much smaller than on the Moon.[1] We see the logic in such a proposal, but we propose that a Lagrange Point space port is a longer-term option than the medium-term prospect of a lunar space port. We further note and concede that at times both Elon Musk and Jeff Bezos were not attracted to refuelling options and a space port role for the Moon. In recent months, however, Elon Musk has shifted SpaceX's near-term priority toward the Moon, which could be achieved in less than 10 years, while Mars remains the long-term goal. Reuters reported on 8–9 February 2026, that Musk said SpaceX had redirected focus to building a "self-growing city" on the Moon, with an uncrewed lunar landing targeted for March 2027 [5]. In this book, we shall hold the likelihood of the Moon being a vital step to future developments to be high (see the research

[1] There are actually five Lagrange points in the Earth-Moon system, there is the Earth-Moon L1 point, which lies on a straight line directly connecting the Earth and the Moon. At a Lagrange point, the gravitational forces from the two bodies (the Earth and the Moon) exactly balance. Generally, space equipment placed at a Lagrange point might be expected to hold its position easily as the gravitational forces are so low. The L1 point, however, is not stable and that is even before thinking of the role of the Sun's gravitational field. There are, however, two stable Lagrange points (L4 and L5)—one ahead and one behind the Moon's orbital position. These would be good places to place a refueling space station where its orbital stability could be maintained with relatively little thruster fuel use.

notes in Chap. 8). We do not consider direct-to-Mars or Lagrange Point space station scenarios any further in this study.

The Moon's lack of atmosphere removes the need for launch vehicles to use staged sea-level and vacuum optimised engines. This reduces the mechanical complexity of launch vehicles and facilitates the development of single-stage launch technology further reducing technical complexity. Thus, such a launch facility could function as a gateway for Mars, the asteroid belt and beyond, making space exploration less complex and significantly more cost-effective. The presence of lunar ice would also facilitate the development of an independent, permanent human settlement which could function as a seed vault and insurance policy for long-term human survival. In the event of a catastrophic event on Earth caused by a natural disaster, pandemic, or nuclear war, such a colony could serve as a refuge beyond Earth and as a base for repopulating the planet, reducing the likelihood of human extinction.

While Elon Musk has done much to stress the importance of humanity becoming an interplanetary species, he is not the originator of such ideas, nor even the first populariser. For example, Stephen Hawking repeatedly stressed, from 2008 onwards the importance of space exploration for long-term survival. In April 2008, commenting at an event to celebrate the 50th anniversary of the founding of NASA, he is reported to have said regarding humanity spreading out into space: "It will completely change the future of the human race, and maybe determine whether we have any future at all" [6] Probably the first person to imagine humanity as an interplanetary species was Russian space pioneer Konstantin Tsiolkovsky, who, before the first world war, memorably observed: "Earth is the cradle of humanity, but one cannot live in the cradle forever." [7].

In the 2020s, the idea that humanity must diversify away from planet Earth has taken on an urgency. In part, this has been driven by the impact of the COVID-19 pandemic, which spread around the world far faster than contagious diseases in previous centuries. The threat of global nuclear war appears to be growing and we are increasingly conscious of our planet's vulnerability to a devastating asteroid impact. One might joke that the chance of something dreadful wiping out life on planet Earth is only about one in a million each year, but when one realises that there are around 1000 such risks all in play at the same time and the list of threats is growing fast, one starts to think we need a lifeboat for our civilisation. It is such thinking that, to a large measure, motivated Elon Musk's passion for Mars. It is noteworthy that he has now shifted his attention to the Moon. When one compares the challenge of establishing a permanent and sustainable human settlement, Mars is the place where long-term chances are best, but to get from where we are to such a situation is not going to be quick—of the order of at least 20 years. The Moon is a harder place to achieve sustainability—less gravity, and a harsher solar wind environment than is found on the more distant Mars, whose surface is partially protected by the Martian atmosphere. Interestingly, neither planetary body has the benefit of the Earth's magnetic field, which is so vital to the survival of life on Earth. One might argue that Musk turned his attention away from Mars and towards the Moon, not primarily because the Moon was easier, but rather from a fear that the existential dangers to the Earth are now so proximate that Mars can only become viable after it is too late. The reality, however, appears to be rather more prosaic and practical—namely that SpaceX was awarded a major contract by NASA for the Artemis programme. SpaceX is to provide the Human Landing System (HLS), which is the spacecraft that will take the astronauts from lunar orbit down to the Moon's surface and back.

Whatever the role of interplanetary species thinking in actual space strategy and policy, it is clear that if humanity is to achieve that status of an interplanetary species, then energy supply at the level of MW would appear, to us at least, to be a prerequisite for true, i.e. enduring, success.

Economic considerations further strengthen the case for returning to the Moon. The lunar surface is expected to contain valuable resources, including helium-3, a potential fuel for fusion energy, and rare earth metals which are essential for modern technology. Extracting these elements could help accelerate the energy transition while reducing the dependence on terrestrial mining which is often highly damaging to local ecosystems. This makes the Moon attractive for both industrial and scientific ventures and has the potential to revolutionise the Earth's economy, should it prove viable.

Finally, it is impossible to discuss the renewed interest in lunar exploration without acknowledging geopolitical competition and national agendas. Just as the original space race between the United States and the Soviet Union was fuelled by the desire to demonstrate technological superiority, a new era of space competition is unfolding, particularly between the United States and China. China has made significant strides in its lunar programme during the last two decades and currently accounts for four of the nine successful soft landings on the Moon in the past 40 years [2]. This new dynamic in space exploration may spark a second space race, this time with the goal of establishing a permanent human settlement on the Moon.

The UK has in the mid-2020s started to talk about British interests and opportunities concerning the Moon. In particular, in November 2025, the lunar Policy Platform and the Royal Astronomical Society (RAS), in collaboration with the UK Space Agency (UKSA) organised a UK Lunar Community Workshop. The workshop was held at the London premises of the Royal Astronomical Society under the title: *The UK on the Moon—opportunities, challenges and the road ahead*. According to a report by Keerthi Gopalakrishnan, the consensus of the meeting was that the British position should be that access to the Moon be global, open, and collaborative [8]. Governance issues featured prominently with a particular emphasis on the need for planetary protection of the lunar environment. There was an emphasis on science and discovery best undertaken in collaboration with other nations. Gopalakrishnan reports that there was a sense of scepticism around the commercial prospects for the near-term commercial extraction of lunar materials, including helium-3 for return to Earth. There was a sense that in situ resource utilisation (ISRU) is likely to be developed. The meeting noted that space tourism showed the greatest potential of all the commercial opportunities likely to drive progress.

As the authors of a book entitled *Fusion on the Moon*, we make the following observations in response. First, we concur with the logic as reported, but we see it constrained to an admittedly orthodox scenario in which energy is in very limited supply, i.e. perhaps because there is no fusion on the Moon. We have (in Chap. 5) suggested that space tourism has a viable, but limited, future without MW energy supply. Second, we observe that the UK is not a leading nation in the human aspects of space exploration. For decades, UK scientific policy has favoured robotic missions over human space flight. We are not here saying that the UK's historical aversion to human space exploration was a bad decision, all we are saying is that in the area of human space flight, the UK has been a follower, not a leader. That cannot be said for fusion energy, where the UK is most definitely a leader. The UK was home to the world's largest tokamak experiment—the Joint European Torus (JET), which operated for 40 years 1983–2023. Today, the UK is home to an emerging cluster of private and public fusion energy enterprises [9]. The 2025 *UK on the Moon* workshop called for international partnership and collaboration (see reference [8]). Perhaps in a scenario of fusion on the Moon, the strengths of the UK in fusion science and technology can be traded for partner expertise in human space exploration. One partner country that some of us (WJN and DWW) have identified as a particularly appropriate future collaborating nation in Fusion is Japan. As important high-tech 'middle powers' in the language of the new geopolitics, a UK–Japan collaboration oriented to Fusion on the Moon could be very exciting indeed.

Text Box 1: Artemis and the Hunt for Purpose in American Space Policy

The US-led Artemis programme encompasses a structured series of missions and the associated Artemis Accords. It represents a strategic drive to reestablish a US-led sustained human presence on the lunar surface. However, the programme recently underwent its most significant overhaul to date. As of 27 February 2026, NASA officially decoupled the Artemis III mission from its goal of a lunar landing [10]. Artemis III is now slated for mid-2027 as a crewed test mission in low Earth orbit (LEO) to validate docking and life support systems. The first crewed lunar landing since Apollo is now reassigned to the new Artemis IV mission, targeted for early 2028 [10].

The mission component has encountered notable delays. While the Apollo programme was a model of complex systems engineering, Artemis has struggled with modern hardware reliability. Boeing, in particular, has faced severe criticism. In February 2026, NASA officially designated the 2024 Boeing Starliner crewed test flight a 'Type-A mishap'—the most serious level of non-fatal incident—citing 'cultural breakdowns' and 'qualification gaps' [11]. Consequently, the next Starliner flight originally scheduled for April 2026 has been delayed (as of August 2026) and downgraded to an uncrewed cargo mission. Additionally, the Artemis II lunar flyby was recently delayed from March to 1 April 2026, due to a dislodged seal in the rocket's upper stage helium system [12]. The mission successfully passed around the Moon on 10 April 2026.

Despite these setbacks, the conceptualisation of the Moon as a testing ground remains a central tenet of US policy. This objective does not preclude the development of fusion power for the lunar surface. The Artemis Accords, while creating friction with partners like China and Russia due to their perceived US-centricity, provide a framework for resource utilisation that could eventually govern activities like helium-3 extraction for fusion.

1.3 The Outer Space Treaty and the Lunar Agreement

The Outer Space Treaty formally known as the *Treaty on Principles Governing the Activities of States in the Exploration and Use of Outer Space, including the Moon and Other Celestial Bodies* establishes outer space as a formal international space in legal terms [13]. The Treaty was established in 1967, and it has been supported by all major nations involved in space exploration. The Treaty establishes outer space, including the Moon, as a place of peace with access to all. The deployment of nuclear weapons is specifically prohibited as are explicitly military activities. The Treaty does not prohibit defensive capabilities in principle. The Treaty takes its logic from the Antarctic Treaty (see Text Box 3 in Chap. 2) established 10 years earlier. It has been argued [14] that US enthusiasm for its Artemis Accords (see Text Box 1) is an attempt by the US to render the Outer Space Treaty irrelevant. The Outer Space Treaty created an international legal framework in which subsequent agreements have gained traction including, for example, in space rescue protocols [15]. One such related initiative has been Lunar Agreement [16]. It was developed via the United Nations Office for Outer Space Affairs and was adopted by the UN General Assembly under resolution 34/68 on 5 September 1979 [16]. The Lunar Agreement initially received broad support, but in some places attitudes soon hardened and today the Moon Treaty is not supported by any of the major space powers and most notably not by the United States.

1.4 The Case for Lunar Fusion

Achieving sustained fusion reactions for power generation on Earth—let alone the Moon—is a generational challenge, requiring significant advancements in plasma physics and engineering.

In Sect. 3.1, we will outline the various possible fusion reactions that might be deployed for useful energy generation in a lunar context. Each has different strengths and weaknesses. The plasma physics issues will be described there, but for now, it is sufficient to appreciate that deuterium-tritium (D–T) fuelled fusion is the front-runner technology for terrestrial use. It yields a very fast (14 MeV) neutron. It is that neutron that removes most of the useful energy from a D–T fusion device. These neutrons are emitted in random directions and they pass through the first wall of the plasma vessel depositing their energy in a 'blanket'. From there, the heat of the blanket can be used to power an electricity-generating turbine. These neutrons are also key to the fuel security of the DT fusion proposition. The proposition is to multiply the number of neutrons in a breeder blanket and to use the resulting neutrons to produce tritium fuel from lithium-6. The neutrons from D–T fusion will cause microstructural damage to components of the fusion device, and much care must be taken in design to mitigate such issues. Lastly, these neutrons also cause the vessel of the tokamak to become radioactive although this is likely to be substantially less important to space agencies or other actors than spent fuel from fission sources. Adapting fusion technology to the lunar environment presents both advantages and challenges.

Importantly, the lunar vacuum could simplify vacuum chamber requirements, permitting simple mechanical venting to the ultra-high vacuum (UHV) of the Moon. Furthermore, the cryogenic temperatures of PSRs could potentially aid in cooling superconducting magnets. Of course, despite these possible advantages, the construction, maintenance, and operation of a complex fusion reactor on the Moon remains a substantial engineering proposition.

A comparative analysis might suggest that fission power offers a more immediate and predictable solution for lunar power generation. However, this book does not consider this to be an enduring truth, as we posit that fusion will overtake fission as a technological choice. Any early fusion reactor on the Moon is likely to share the lunar surface with fission reactors, whether co-located, or elsewhere. Fusion power holds the promise of a more sustainable and abundant long-term energy source, particularly if helium-3 mining and local use proves feasible. Environmental considerations are also crucial. Fission produces long-lived radioactive waste even if sealed-for-life cores are used. In this book, we suggest that as lunar settlement expands, the policy pressure to avoid leaving radioactive wastes on the Moon will grow. We suggest that the threshold for exemption might be such that fusion is regarded as compliant, but that future fission technologies are barred or are subject to very onerous decommissioning requirements.

In addition, there are substantive benefits to operating a fusion reactor on the lunar surface that deserve serious consideration within a 2040–2070 timeframe, as shown in Fig. 1.2. These benefits are drawn out of the text in detail in the following sections, but in short, a lunar D–^{3}He fuelled reactor has no need of a breeder blanket (reducing mass, and therefore thermal cycling stresses), it has no rotating turbomachinery (and therefore is more reliable in a remote, lunar context), it receives a cryogenic boost from the natural temperature of its environment, it has access to atmospheric vent without vacuum plant, and it has ready access to a supply of ^{3}He without the need for any artificial fuel manufacture. The natural abundance of ^{3}He on the Moon avoids a bottleneck for terrestrial D–^{3}He reactors. Of course, terrestrial D–^{3}He reactors may also be fuelled by helium-3 from the Moon. That is a scenario we shall later term '^{3}He for export'. In that scenario, ^{3}He use on the Moon is greatly favoured. Furthermore, if a lunar ^{3}He export industry were to emerge, then terrestrial D–^{3}He fusion would surely become viable despite the fusion condition challenges explained in Sect. 3.1. With such ideas in mind, we see D–^{3}He scenarios where fusion on Earth might even come after fusion on the Moon (compare second and fourth coloured block columns in Fig. 1.2).

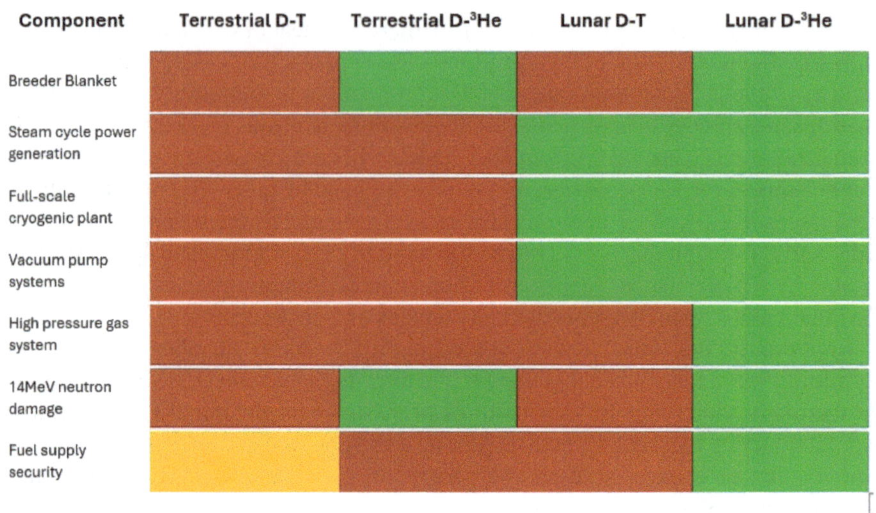

Fig. 1.2 System analysis for lunar and terrestrial fusion reactors—red denotes a challenging technical requirement, whereas green indicates that the requirement is avoided or trivially satisfied. The wider challenges of the lunar location and logistics are here assumed to be manageable. The amber cell is discussed in the main text. Authors' own work

Figure 1.2 adopts a 'traffic light' colour scheme in considering possible fuel cycles and their infrastructure implications for fusion on the Moon. Red implies a significant challenge, whereas green indicates that the technological requirements can be avoided or easily mitigated. Terrestrial D–T fuel security is shown as amber noting the severe neutron economy and fuel conversion challenges in that case. Of course, significant effort is underway to establish sustainable fuel production for D-T fusion on Earth. Green, however, is not chosen for that cell, as we take the view that terrestrial tritium generation will be extremely difficult challenge for an emergent fusion industry. These challenges are, of course, nothing per-se to do with fusion on the Moon. Fuel security for terrestrial D–^{3}He fusion is shown as red, as this acknowledges the difficulty of sourcing sufficient ^{3}He on Earth.

1.5 Power Requirements on the Moon

The feasibility of deploying fusion power on the Moon must be assessed within the broader context of energy demands for human operations on the lunar surface. Given the immense technical challenges associated with developing and deploying a functional fusion reactor on the lunar surface, such an endeavour would only be justifiable once the power requirements of lunar operations surpass a certain threshold. To determine when this might occur, Fig. 1.3 outlines the estimated power demands

Activity (increasing complexity)	Power requirements	Constant powe required
Robotic exploration	0.1-1kW [16, 19]	Sometimes
Robotic exploration with sample return		Sometimes
Short duration human exploration	1-10kW [20]	No
Periodically occupied human outpost	10-200kW [17]	No
Permanently occupied human outpost		Yes
Space port	1-10MW	Yes
Mining operations in support of Lunar needs		Yes
Mining operations in support of Mars needs		Yes
Mining operation in support of Earth's needs		Yes

Fig. 1.3 Power considerations associated with different levels and type of lunar activity, based on third-party literature as estimated and based on Authors' estimates. Scenarios for the future of space tourism can emerge at a range of power levels, as described in the main text. Authors' own work

for various types of lunar activities in increasing difficulty, ranging from previously accomplished missions to highly complex future operations.

Activities highlighted in dark blue at the top of the table represent mission types that have already been successfully demonstrated and serve as useful benchmarks for estimating the current power requirements on the Moon. These include historical programmes such as the Apollo, Luna, and Surveyor missions, as well as more modern operations such as the Chang'e missions, Chandrayaan 3, and Blue Ghost 1 [17]. Most of these missions were limited to a single lunar day and relied on a combination of solar panels and pre-charged batteries for power. The few missions that lasted longer than a single lunar day operated in power-saving during the 14 day lunar night, using battery reserves recharged during daylight hours. This demonstrates that current lunar activities have no need for fusion power as they are sufficiently powered by existing solar and battery technologies.

Activities highlighted in the middle of the table represent mission types that have yet to be conducted on the Moon but have been conducted within a space context. These represent activities that may need fusion power to operate on the Moon, although it is likely that better alternatives exist. Examples include the International Space Station (ISS) and Tiangong space station, both of which currently serve as solar-powered human outposts in space. The ISS currently demonstrates humanity's ability to sustain relatively high energy demands in space. The ISS can generate an average 105 kW of electrical power [18]. However, replicating similar power levels for continuous operations on the Moon presents additional challenges, particularly for power-intensive activities that must endure the lunar night. A permanently occupied outpost would require reliable, uninterrupted energy for life support and scientific experiments, making it difficult to depend solely on solar power and battery storage. This storage problem may even limit the viability of solar power for an outpost which is only occupied during the lunar day and hibernates during the lunar night depending on how much the power requirements can be reduced during such a hibernation. While fusion power might eventually offer a robust solution, it is more likely that a hybrid approach combining solar and fission energy would suffice. Under this scenario, solar panels could provide ample energy during the lunar day, while a modest fission-based reactor (e.g. similar to the 3 kW BES-5 reactors that were used in Soviet spy satellites [19]) would ensure that critical systems remain operational throughout the lunar night.

Activities highlighted in the lightest blue at the bottom of Fig. 1.3 represent mission types that have not yet been attempted in any space context. These activities, including resource extraction, manufacturing and the permanent human settlements needed to operate them, demand a continuous and substantial power supply far beyond the capabilities of current space power sources. Furthermore, the requirements for constant, high levels of power make solar power and battery solutions impractical, making the cells shown in Fig. 1.3 promising candidates for fusion power on the Moon. However, these high-power operations are also the most technically challenging, requiring decades of development and immense financial investment. Given these hurdles, there is no certainty that lunar activities will ever reach this level, making the eventual necessity of fusion power on the Moon uncertain.

We have argued previously about the attractiveness of PSR locations that are very cold and completely dark. While lighting is associated with a very minor power requirement, heating and the use of industrial machinery in an ambient cryogenic (circa 50 K) context could be very energy intensive. We note that energy losses on the Moon are purely radiative and conductive, unlike on Earth where the atmosphere provides a third mode of energy dispersion and hence waste. Nevertheless, we posit a need for MW scale provision for industrial activity and associated settlement in PSR. Our claim is, however, something that we suggest requires further research, as we believe it is a key driver for the case for fusion on the Moon.

We should stress at this point that although fusion on Earth is often discussed as a GW technology that is very far from our thinking for the Moon. A tokamak concept for the Moon will be very different—even if D–T fuelled, it will not generate its own fuel, simply because the upmass concerns around in situ lunar breeder blankets are relatively daunting.[2] We are not saying that heavy lift capabilities are insufficient to deliver breeder blanket modules; rather that it is a technical challenge best avoided

[2] We note that generally term 'upmass' is a space logistics term used to describe payload mass lifted to Earth orbit. We use it, by extension, to denote payload mass lifted from the Earth to the Moon

either by providing tritium from Earth (a tiny upmass concern) or by moving to ^{3}He fuel cycles.

1.6 A Key Competitor—Nuclear Fission on the Moon

As we examine energy needs for lunar activity, we cannot neglect to consider a key alternative: nuclear fission. We suggest that for a concentrated power source at the heat of a settlement 'citadel' or major industrial facility, it is the only reasonable competitor in terms of power density. It also has substantial benefits over non-nuclear power generation at MW scale. Finally, only nuclear options would appear viable as a local energy source for cold dark lunar polar contexts. As with fusion, nuclear fission energy relies on the application of Einstein's equation:

$$E = Mc^2$$

In this case, however, instead of two light nuclei being fused together and releasing energy, a heavy nucleus is split. This results in the formation of two, or more, lighter nuclei and the release of one, or more, neutrons. As the mass of the nuclei formed and neutrons resulting from the fission is less than that of the original nucleus, energy is also released. In most cases, this fission requires the original nuclei to be struck by a neutron. Thus, the neutrons released during fission can go on to cause further fissions. As more than one neutron, on average, is released during fission and losses due to absorption and leakage are sufficiently low, then a 'chain reaction' can be established and useful energy obtained. Typically, a uranium fission event releases, on average, 2.4 neutrons. This is better, in terms of neutron economy, than the one neutron emitted by D–T fusion. Each process consumes one neutron. In the case of D–T fusion, that neutron is needed to convert ^{6}Li to tritium.[3]

It is likely that there will be an initial phase of fission's use in lunar exploration and settlement. However, we here explore the attributes of fission on the Moon and conclude that it may be blocked as a long-term sustained option.

A major disadvantage of nuclear fission is that many of the residual nuclei from the fission process (i.e. 'fission products' and 'minor actinides') are radioactive and are therefore potentially hazardous. In contrast, the nuclear fusion process does not produce such products. Both approaches to nuclear energy do, however, produce nuclear waste via neutron activation of reactor components.

---

and we use the term deliberately to remind that every kg lifted from the Earth to the Moon is an economic cost best avoided, if possible.

[3] It is planned that D-T fusion systems will make use of neutron multipliers (e.g. based on beryllium or lead) to convert a high energy neutron into a larger number of neutrons.

1.6.1 Prior Studies

Petrocelli and colleagues [22] have surveyed the history of programmes to design nuclear reactors (and other radiological energy sources) for employment on the Moon and in space more generally. Currently, NASA is funding three programmes to develop fission reactor design concepts for use on the Moon. The companies funded are Westinghouse [23], Lockheed Martin [24] and IX (a joint venture of Intuitive Machines and X Energy) [25]. Similarly, a joint Russian–Chinese programme has been announced [26]. In the United Kingdom, Rolls-Royce have received Government funding to develop a lunar reactor design concept [27].

Details of these design concepts are sparse, but it is likely that they will be 'fast' reactors. The fast reactor approach has the advantage that there will be no need to incorporate a moderator to slow neutrons. A moderator increases the likelihood that fission will occur. Most, indeed nearly all, commercial nuclear reactors on Earth use a moderator, but this comes at the expense of a much larger reactor core. Typically that consequence has not been a major concern for power station projects. The absence of the need for a moderator for a lunar reactor would reduce the complexity of the designs and is likely to reduce the mass that needs to be transported to the Moon.

1.6.2 Fission Fuel Supply

The use of fission reactors on the Moon would bring with it the need to transport fission fuel to the Moon. This fuel is composed of uranium and/or plutonium, both of which are radiotoxic (to varying degrees) and are heavy metal poisons. The launch of a rocket into space carries with it the risk of failure and the destruction of the payload. For example, the January 2025 failure of a Space-X rocket launch resulted in debris falling on the Turk and Caicos Islands [28]. Lenkart and Biller report [29] that since 1970, the average annual rate of failure of rocket launches is 6.6%. Any failure of a launch that includes fission nuclear fuel as part of its payload therefore carries a significant risk in principle, at least, of radioactive material dispersal into the environment. The safety issues around the escape of tritium from a fusion based alternative are generally less troubling, but they are and not entirely negligible. On balance, the issues are less problematic than in the case of a fission reactor related accident. In the event of an accident within the atmosphere associated with fusion fuel tritium, the majority of the emitted tritium would rapidly disperse and rise high in the atmosphere. It would also decay relatively rapidly give its 12 year half-life. It should be noted that in confined spaces, such as a building, tritium can be a major health hazard, but that situation does not generally apply to space system accidents.

Any nuclear reactor on the Moon, whether a fission or fusion reactor, has the potential to undergo an accident that results in the release of radioactive material into the lunar environment. In the case of a fusion reactor, this will be a relatively small amount of tritium and perhaps some activated parts of the fusion reactor. These parts might have been activated by the neutron flux in use or have absorbed radioactive tritium into the reactor materials. Tritium is a form of hydrogen and it penetrates metals relatively easily, for example. For a fission reactor, a large amount of fission products can be expected to be released, although some fuel forms can be designed with accident safety in mind. Also, the energy behind the release can be expected to potentially larger in the fission case, but we do not wish to be alarmist about fission safety. Issues that appear problematic in principle are, in fact, easily managed through good design and engineering. But, for the sake of argument, let us imagine a worst-case fission accident on the Moon. Many of the nuclides released will be volatile species (e.g. Iodine) which may condense in the cold environment and be deposited on the lunar surface. In contrast, fusion tritium would immediately dissipate into the vacuum of space—where it would later decay away relatively quickly.

1.6.3 Fission Waste Return and Space Safety

The issue of whether, or not, it will be necessary, perhaps because of international agreements, to return radioactive (and other) wastes to Earth is a complicated concern that largely lies beyond the scope of this book—we only remark that it is possible that new requirements for lunar surface actors to decommission their sites may come into force. Indeed we, as authors, expect a tightening, rather than a weakening, of the regulatory regime around wastes and decommissioning, if the lunar development story is a success.

If, in the future, it is necessary to return the radioactive wastes to Earth this would introduce additional hazards. For a fission reactor, the spent nuclear fuel will form most of the waste (at least in terms of radioactivity). It is important to remember that a used fission fuel core will represent a vastly larger radioactive hazard than the fresh reactor fuel that would originally have been launched from Earth. The build-up of fission products and minor actinides in the core during operations ensures that an Earth return scenario for fission will never be easy. Such ideas link to the hazard that the launch of the rocket returning to Earth will fail leading to the distribution of the spent fuel on the lunar surface. Secondly, there are the hazards associated with reentry to the Earth's atmosphere and, if reentry itself is successful, a failure to make a controlled landing at the planned landing site. Each of these failures has the potential to result in the distribution of the spent fuel throughout the lunar or terrestrial environment. This would be a huge setback for the space community. The nuclear community, in contrast, is more familiar with such politics. Data on the proportion of return journeys that will experience such failures is sparse but it likely to be non-negligible.

Issues of flight safety around radioactive materials are with us today. For example, widely used actinide-based Radioisotopic Thermoelectric Generator (RTG) technologies are compact and power dense. They must be protected against catastrophic failure on launch. The use of RTG and similar technologies is a familiar concern for space mission planners. We note, for example, the contentious use of plutonium-238 fuelled RTGs in the NASA Cassini series of missions.

One option for lunar fission wastes is, of course, In Situ Lunar Radioactive Waste Disposal. If the disposal of radioactive waste (as opposed to simply leaving it in place) on the lunar surface is permitted, then presence of spent fission fuel will greatly increase the amount of work that will be needed to dispose of the waste in an engineered facility on the Moon. For a near-surface or deep radioactive waste disposal facility on Earth, the major pathways by which radioactive material can be returned to the surface in an uncontrolled manner are inadvertent human intrusion or intrusion by other species and the return of dissolved radionuclides to the biosphere via groundwater. None of these pathways are relevant in the lunar environment. Therefore, the engineering of any disposal facility on the Moon will be simpler than that for a similar facility on Earth, but that does not mean that it will be any more socially acceptable than it has been on Earth.

In contrast, it is not unimaginable that the vast majority of tokamak fusion radioactive waste can simply be vented to deep space simply by opening a large flange. It is much easier to evaporate small amounts of light atoms to the vacuum of space. Such options do not exist for hundreds of kilograms of highly active heavy actinide waste from fission.

Finally, one might imagine that rather than send fission wastes back to Earth one might send it off into deep space. Such a plan would require the reactor core, or waste package, to escape the gravity well of the combined Earth-Moon system. That is a major technical challenge. Any failure to achieve escape velocity could place the waste package into a hazardous Earth (or Moon) crossing orbit with very significant approach velocities, especially if the orbit is highly eccentric.

In summary, we posit that there is no easily viable deep decommissioning option for fission on the Moon. One might consider fast reactor-based waste incineration and minimisation, but we suggest that would really make fusion on the Moon look relatively easy.

While unsustainable practices might be tolerated in the current era of exploration and in the near future of adventurous resource exploitation, we find it hard to imagine such practices will sit well in the long-term alongside sustained settlement with semi-autonomous communities on the Moon. Such settlements will have children, schools and hospitals, etc. For such a community, higher standards of sustainability will be expected, indeed demanded. We posit that by the end of the twenty-first century—the delivery of new fission systems to the Moon might be banned. Such a scenario prompts us to look across again at lessons from Antarctica. While fission in Antarctica is not actually banned, fusion in Antarctica would be much more attractive to any politically minded decision maker. We therefore also posit that while fission on the Moon is technically attractive in the short term, any near-term success should not deflect from efforts to transition towards fusion on the Moon.

1.6.4 Near-Term Fission Options

Fission power offers a degree of predictability and reduced developmental risk—decades of terrestrial operational experience provide a solid foundation for adapting fission reactors to the lunar environment. Notably, the ongoing development of small modular reactors (SMRs) aligns well with the logistical constraints of lunar deployment, potentially simplifying transportation and installation, with self-contained concepts such as the Westinghouse eVinci currently proving out the kind of autonomous and mechanically simple systems needed for lunar longevity.

As regards the supply of a fission power system from the Earth: that is a familiar and manageable logistical burden. It will be controversial, but it is doable. Operational fission reactors have already been deployed in space. The USSR deployed more than thirty 2 kW BES-5 nuclear reactors in support of its Radar Ocean Reconnaissance Satellite (RORSAT) programme which ran from 1967 to 1988 as a series of missions code-named Cosmos. This was followed by the slightly higher power rated Topaz-1 fission reactor of 1987 which was used on Cosmos missions 1818 and 1867. Each Topaz reactor operated for several months. A successor design Topaz-II has been publicly evaluated by US Los Alamos national laboratory experts for lunar applications [30]. The considerations in that analysis mirror many considerations adopted in this book for our thinking around Fusion on the Moon. The authors of the study, Polansky and Houts saw no major technical obstacle in the way of US utilisation of the Russian fission technology.

1.6.5 Summary of Fission Issues

Noting that radioactive risks from launch failure are manageable and that appropriate safety measures have been developed [31], we posit that the greatest obstacles to fission on the Moon are social and political. The fact that the barriers are not physics and engineering does not mean they are not very real. We note again the prohibitions in the Antarctic Treaty (1959) against radioactive waste. We note that very real difficulties encountered in numerous countries in seeking to establish the geological disposal of radioactive wastes. We posit that the pressure to avoid fission waste and to ensure complete 'actinide decommissioning' could be unavoidable. We posit that total actinide removal from the lunar environment is such a challenging requirement that it might be preferable, at some point in the future, to turn away from fission and towards fusion, even if it has not already shown itself to be surprisingly easy and useful, as this book tends to argue might be the case.

In summary, we take the following view concerning long-term policy options for fission fuel:

- *It cannot be retained indefinitely on the Moon.*

- *It cannot be blasted off into deep* space—the *kinetic energy requirement is too large.*

- *Furthermore, any risks around the ejected item becoming an Earth-crossing object must be mitigated and managed.*

- *The fuel core might be returned to Earth, but at some significant cost and risk.*

- *The use of a small fast reactor on the Moon might reduce, but would not eliminate, the actinide and fission product waste management challenge.*

- *With regard to fission nuclear fuel reprocessing on the Moon, for vitrified* high-level *waste return to Earth; we expect this to be a significantly more difficult proposition than any fusion ideas considered in this book.*

In this book, we do not give special consideration to fusion-driven sub-critical fission reactors or to fusion–fission hybrids. We regard the associated decommissioning issues as being equivalent to a conventional fission reactor.

1.7 Meet High Lunar Electricity Demand with Renewable Energy?

Let us consider the use of solar photovoltaic energy in two contexts—lunar polar PSR and lunar equatorial locations. In this book, we emphasise PSR locations (see Sect. 4.1) and hence let us first consider the use of renewable energy in such a situation. Clearly there is no wind, there is no biomass and there is believed to be no prospect for the equivalent of geothermal energy on the Moon. The only candidate renewable energy source is sunlight. Issues around the solar illumination of the lunar poles takes us to the question—*Are there seasons on the Moon?*

Interestingly the answer is yes, but the difference between the lunar summer and the lunar winter is very mild and that is because the axial tilt of the Moon's rotation (at 1.5°) is much smaller than the equivalent tilt for the Earth (at 23.5°). Planning for the use of solar PV at lunar polar locations cannot completely ignore the issues of the lunar winter—as shown in Fig. 1.4.

We posit that the lunar poles could soon become highly contested spaces especially where permanently dark craters and permanently bright ridges at crater rims can be found. If solar power is the primary, or only, energy source, then the competition for bright ridges could be intense. Furthermore, the sunlight coming at a shallow angle from near the horizon will give rise to great risks of mutual shadowing if two competing industrial interests get too close to one another.[4]

[4] Such concerns link to two important concepts from the Outer Space Treaty—namely 'harmful interference' and 'due regard' (article IX). Article IX has, in turn, prompted consideration of 'safety zones' in the Artemis Accords, for example.

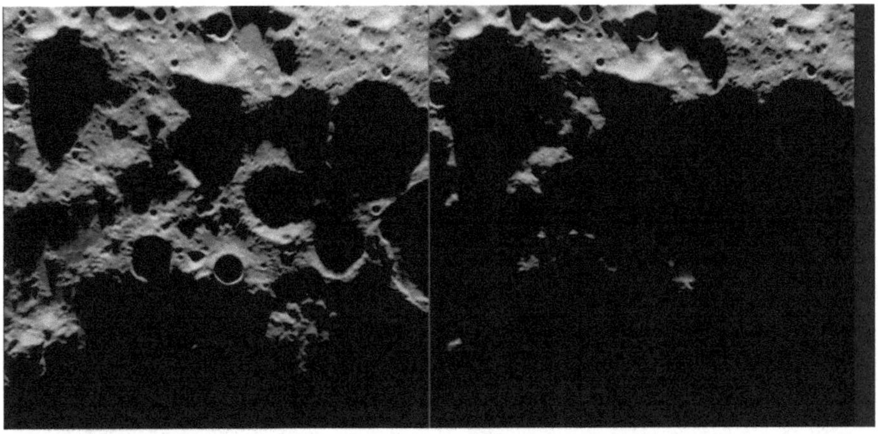

Fig. 1.4 The south pole of the Moon in summer (left) and winter (right). Solar PV resources can be expected to be very precious, and hence contested, in the winter. *Source* NASA [32]

We suggest that there is no easy way in such a challenging location to provide baseload firm power at the MW scale, but that does not mean that MW power is not available for short periods. We do not want to write off the renewable option too easily.

For example, one might imagine 50 kw of solar photovoltaic (PV) running continuously on a bright mountaintop 1 km away from dark mine workings. Such a continuous power source might charge batteries and supercapacitors over several days sufficient to permit short periods of high-power operations—enough perhaps to use sintering or bedrock melting to seal cracked tunnel surfaces to render them safe for pressurisation and occupation. The alternative to sintering or melting would be to use a chemical binding agent lifted from Earth as a sealant. If sufficient lunar water is available to mix with such a binder and with lunar regolith, then such might also facilitate construction with only modest power requirements [33]. However, we are not saying definitively that a MW power source is needed; we merely posit it could help de-risk some very difficult challenges.

Let us briefly step aside from consideration of polar activities and consider the energy needs of a large human settlement on the Moon. For various reasons including solar irradiations and ambient temperature, such a location might be near the equator. Here, the long lunar night (approx. 14 Earth days) makes continuous renewables-based electricity supply at the MW scale difficult. Discounting the use of high capability battery systems for upmass reasons, one might imagine a series of equatorial solar PV power stations spaced around the planetoid in such a way that the sun never sets on the power system. Such a system might deploy cables to connect the power sources, and this might operate as HVDC, or perhaps one might even use microwave power transmission (MWPT) technologies. Another competing idea that has emerged recently is space-based solar power. Specifically we are here thinking of solar photovoltaic satellites in a near-stable lunar orbit (a challenging goal in itself) or perhaps at a Lagrange Point (see Sect. 1.2). One ambition shared with the previous proposition would be to achieve permanent, or nearly permanent, solar irradiation. The resulting electrical power would be transmitted wirelessly to receivers on the lunar surface by MWPT or laser light. For further information see [34, 35]. Such ideas raise many challenges.

Again, we are not saying that non-nuclear options are impossible; we are however noting that they are not obviously easier than fusion on the Moon.

1.8 Project Assumptions, Scope, and Boundaries

This book seeks to explore different commercial scenarios and technological opportunities associated with the use of nuclear fusion systems on the Moon. It is open-minded to all fusion-related ideas, although the greatest consideration is given to the potential application of tokamak devices on the lunar surface. For this study, defence-relevant applications of a lunar tokamak are deliberately excluded from the scope. This exclusion does not imply that future military activities on the Moon are improbable; rather that such issues are not relevant for this book.[5] Nevertheless, the influence of geopolitical competition on shaping lunar activities is acknowledged as a relevant factor driving wider trends in lunar activity.

Throughout this book, it is assumed that humanity will eventually establish a sustained presence on the Moon. This assumption is supported by several key points. The Moon is Earth's closest celestial neighbour and therefore offers an ideal location for testing technologies intended for the exploration of other celestial bodies. Such testing is crucial for de-risking the human exploration mission, since the Moon's proximity to Earth allows for rapid rescue and has negligible radio-lag, compared to Mars.

1.9 Three Key Lunar Resources

In the next section, we shall start to unpack the various drivers that could act in favour of fusion on the Moon. We hope and trust that it is already clear from the context presented above that we see three lunar resources of being critical to future lunar developments including ambitious ideas like fusion energy.

The three key resources are the following:

Water.
Helium-3.
Metals.

[5] We concede that there are an important set of scenarios that we do not consider directly. Namely that the future of technology development on the Moon might be led by military actors. Within such scenarios energy sources might be developed for military reasons and then later civilian adoption might be favoured by the prior military developments. Such issues represent an alternative future to the one considered in this book.

We posit that a deep polar crater on the Moon could provide all three resources in close proximity, and we shall return to that idea in Chap. 5. Figure 1.3 did not discuss lunar tourism as a lunar development requiring energy services. That is because, as we shall argue below, we judge it most likely that the scale of lunar tourism will be a consequence of the energy available rather than a driver, by itself, of energy innovations at the MW scale. We accept that this is an assumption on our part that should be subject to further research and assessment.

1.10 The Allure of Lunar Tourism

Space tourism has long featured prominently in science fiction and popular culture, Films such as Stanley Kubrik's 2001 and more recently The Fifth Element and the animated comedy Wall-E have all made reference to the idea of a vacation in space. Indeed, the first ever science fiction movie *The Trip to the Moon* (originally *Le Voyage dans la Lune*) while ostensibly depicting a scientific expedition, it actually resembled an excursion, or fun day out, for some Victorian gentlemen, at least until things start to go wrong for them.

To date and in real life, most space tourism propositions have focussed on terrestrial sub-orbital flights, such as the trips offered by Virgin Galactic and Blue Origin. Orbital space tourism is more ambitious and such trips are offered by Space-X using its Dragon capsule and by Axiom Space and Space Adventures. Space Adventures was the pioneer of commercial spaceflight and it has proposed, but not yet developed, a tourist route that could orbit the Moon. Space Adventures has long partnered with Russian state-owned space technology providers. Space Adventures was instrumental in the flight, in 2001, of Dennis Tito—the world's first commercial space traveller.

It is clear that, in the 2020s, space tourism remains a source of fascination for many. We suggest that it will have a role to play in the future of humans on the Moon, but with three key considerations in mind relevant to our central proposition of Fusion on the Moon.

First, we observe that while available energy supply will greatly affect the prospects for tourism on the Moon, the absence of a MW power source would not prevent the emergence of a viable space tourism economy. Second, if lunar polar industrial development occurs with the development of MW energy sources, such as might be provided by fusion on the Moon—this would have a transformative effect on the mostly equatorial lunar tourism business, but only towards the end of period of interest, namely in the 2060s. Third, while we see space tourism having important synergies with fusion on the Moon, we do not see it being a major driver of fusion on the Moon technology development. We base this observation that low energy use lunar tourism could be viable and appropriately challenging within its own constraints. We describe such a low energy development scenario in Text Box 2.

The success of lunar polar industrial activity will be a prerequisite for large-scale (approximately 100 visitors) lunar tourism. In this scenario, the lunar hotels of the equatorial region will need water, energy and potentially fuels supplied from polar sites.

Text Box 2: A Low Energy Development Path to Tourism on the Moon

Noting the scale of commercial adventure offerings to climb Mount Everest or to descend to deep ocean depths we see the prospect in the coming decades of lunar tourism at the scale of approximately 10 visitors at any one time.

The need to avoid heavy energy loads would motivate equatorial regions during the long lunar day. Site selection would be similar to the concerns of those planning the Apollo landings of 60 years ago. The prospect of tourists visiting, or if future governance so ensures—observing from an unintrusive distance, the Apollo programme landing sites would further motivate near-equatorial locations. Visits, like those of the Apollo astronauts, would last for a small number of Earth days.

As we note elsewhere, we see a strong prospect for entirely separate large-scale and energy intensive development at the lunar poles. Tourists would be kept away from the polar regions which we expect to be geopolitically contested territory focussed solely on industrial exploitation.

We see the short lunar day visits by tourists to equatorial regions being supported by solar power. One task that the tourists might help with would be to clear regolith dust deposited on the solar panels following the rocket exhaust effects of the previous group of visitors.

Visitor safety (from micrometeorites and solar radiation) will be a key concern. One mitigation path might be to focus on shorter visits, as Apollo did, or to plan to be based near a natural lava tube cave that, in extremis, could be used as a shelter during a surprise solar storm.

For tourism at the scale envisaged here, we do not see the prospect of major construction or excavation. Such options are restricted, in our thinking, to a longer-term future in which MW scale energy supply becomes available transforming all activity on the Moon including tourism. We note that synergies that can exist between tourism and militarism. For example, a state keen to militarise its presence on the Moon, but also concerned for its international image, might choose to develop lunar tourism so that it might later justify quasi-military capabilities in respect of emergency assistance and rescue.

As noted elsewhere (Chap. 2, Sect. 2.2.2), the providers of space tourism services prefer the term commercial space services. In this book, we are comfortable to use the term space tourism as it avoids confusion with an important separate aspect of commercial space services. In that case, true astronauts are conveyed to space on commercial terms. This emerging, and important business is described in the next section.

1.11 Commercial Astronaut Support

Houston-based Axiom Space has pioneered a model of commercial astronaut support that is distinct from conventional space tourism. Rather than simply selling a leisure experience, Axiom acts as a full-service orbital mission provider, organising training, mission planning, International Space Station (ISS) integration and commercial access to low Earth orbit in partnership with SpaceX and NASA. NASA itself frames these private astronaut missions as part of building a commercial low Earth orbit economy, with crews conducting science and commercial activities rather than undertaking a vacation.

A good example is Axiom Mission 4 (Ax-4), a roughly two-and-a-half-week mission to the ISS that ran in June/July 2025. The crew was commanded by Peggy Whitson, Axiom's Vice President of Human Spaceflight and a former NASA astronaut, and included Shubhanshu Shukla of India, Sławosz Uznański-Wiśniewski of Poland, and Tibor Kapu of Hungary. Ax-4 was described by Axiom as the company's second commercial mission made up of government and ESA-sponsored national astronauts, and by NASA as the fourth private astronaut mission to the ISS.

The Polish case illustrates the model particularly well. Astronaut Sławosz Uznański-Wiśniewski, a scientist and expert in radiation-tolerant designs for space applications, became the second Polish citizen in space after Mirosław Hermaszewski (who travelled to space via a Soviet Union mission in 1978). Unlike India's Ax-4 participation, which the Indian Space Research Organisation (ISRO) procured directly from Axiom Space through a Space Flight Agreement, Poland's IGNIS mission was organised through an agreement between the Polish government and ESA, with Sławosz Uznański-Wiśniewski flying as an ESA project astronaut. Similarly, the Hungarian astronaut Tibor Kapu became the second Hungarian in space after Bertalan Farkas, and his participation has been organised through Hungary's national HUNOR programme working with Axiom Space. The Ax-4 mission was supported not only by the use of Space-X technology, but also with technical services from NASA.[6]

We mention this business model as we see it being of relevance to fusion on the Moon. Private companies can allow national governments to achieve their goals in space, using people and capabilities developed by NASA and other leading space agencies. Ax-4 reportedly carried out around 60 scientific studies and activities representing 31 countries, making it Axiom's most research-intensive mission to date and underlining that the value proposition is as much about sovereign capability, science and industrial development as about prestige or personal experience.

This model shows how public and private roles may combine beyond low Earth orbit in the future. Private firms can package transport, mission operations and specialist expertise, while governments fund participation in order to meet national scientific, technological or strategic goals. In that sense, Axiom provides an early

[6] Axiom Space's (Ax-4) mission used a SpaceX Dragon capsule.

example of a broader pattern: commercial space services and space tourism overlap, but they are not synonymous. The commercial market may be driven less by leisure travel than by governments and institutions purchasing access to infrastructure, skills and operational experience developed with NASA and other leading agencies.

References

1. W.J. Nuttall, S. Konishi, S. Takeda, D. Webbe-Wood, *Commercializing Fusion Energy: How Small Businesses are Transforming Big Science*, 2nd edn. (Porto Press, 2025). ISBN 9781849956048
2. NASA Science, in *Moon Missions*. https://science.nasa.gov/moon/missions/ Accessed Date 24 Mar 2025
3. S. Li, P.G. Lucey, R.E. Milliken, P.O. Hayne, E. Fisher, J.-P. Williams, et al., Direct evidence of surface exposed water ice in the lunar polar regions. Proc. Natl. Acad. Sci. **115**(36), 8907–8912 (2018). https://doi.org/10.1073/pnas.1802345115
4. Tintin.com, in *A Lunar Project: Tintin's rocket*. Available at: https://www.tintin.com/nl/news/6343/a-lunar-project-tintins-rocket. Accessed 07 May 2026
5. Reuters, in *SpaceX Prioritizes Lunar 'Self-growing City' Over Mars Project, Musk Says* (updated 9 Feb 2026). Available from: https://www.reuters.com/science/musk-says-spacex-prioritise-building-self-growing-city-moon-2026-02-08/
6. Space.com, in *Primitive Alien Life May Exist, Stephen Hawking Says* (updated 22 Apr 2008). https://www.space.com/5266-primitive-alien-life-exist-stephen-hawking.html
7. K.E. Tsiolkovsky, in *The Exploration of Cosmic Space by Means of Reaction Devices* (1911). https://archive.org/details/nasa_techdoc_19750021068/page/n1/mode/2up
8. K. Gopalakrishnan, in *The UK on the Moon—Opportunities, Challenges and the Road Ahead* (2026). https://www.aerosociety.com/news/the-uk-on-the-moon-opportunities-challenges-and-the-road-ahead. Royal Aeronautical Society AEROSPACE Insight.
9. W.J. Nuttall, S. Konishi, S. Takeda, D. Webbe-Wood, Chapter 14: Conclusions, in *Commercializing Fusion Energy: How Small Businesses Are Transforming Big Science*, 2nd edn., (Porto Press, 2025). ISBN 9781849956048
10. The Guardian, in *NASA announces Artemis III mission no longer aims to send humans to moon* (2026, Feb 27)
11. NASA Blog, in *NASA Releases Report on Starliner Crewed Flight Test Investigation* (2026, Feb 19)
12. NASA Blog, in *NASA Repairs Upper Stage Helium Flow, Preps Continue Ahead of Rollout* (2026, Mar 3)
13. NASA, in *The ARTEMIS Accords: Principles for Cooperation in the Civil Exploration and Use of the Moon, Mars and Asteroids for Peaceful Purposes* (2020). https://www.nasa.gov/wp-content/uploads/2022/11/Artemis-Accords-signed-13Oct2020.pdf?emrc=67e3b2971e4a7
14. C. Newman, Artemis accords: Why many countries are refusing to sign moon exploration agreement. The Conversation (2020) https://theconversation.com/artemis-accords-why-many-countries-are-refusing-to-sign-moon-exploration-agreement-148134 https://www.unoosa.org/oosa/en/ourwork/spacelaw/treaties/introouterspacetreaty.html
15. United Nations Office for Outer Space Affairs, in *Resolution Adopted by the General Assembly: 2345 (XXII)*. Agreement on the Rescue of Astronauts, the Return of Astronauts and the Return of Objects Launched into Outer Space. https://www.unoosa.org/oosa/en/ourwork/spacelaw/treaties/rescueagreement.html
16. Agreement Governing the Activities of States on the Moon and Other Celestial Bodies, Dec 18, 1979, 1363 UNTS 3, available at: https://www.unoosa.org/oosa/en/ourwork/spacelaw/treaties/moon-agreement.html

17. Firefly Aerospace, in *Blue Ghost*. https://fireflyspace.com/blue-ghost/ Accessed Date 24 Mar 2025
18. E.B. Gietl, E.W. Ghouldston, B.A. Manners, R.A. Delventhal, The electric power system of the international Space Station-a platform for power technology development. NASA (2000) Contract No.: NASA/TM2000-210209. https://ntrs.nasa.gov/api/citations/20000063506/downloads/20000063506.pdf
19. N.H. Whitman, D. Poston, P.R. Mcclure, *Modeling and Analysis of the BUK/BES-5 Fast Reactor Using MCNP* (Los Alamos National Laboratory (LANL), 2015) https://www.osti.gov/servlets/purl/1207760/
20. NASA Science, in *Perseverance Rover Components*. https://science.nasa.gov/mission/mars-2020-perseverance/rover-components/. Accessed Date 24 Mar 2024
21. A.B. Campos, Apollo experience report-lunar vXodule electrical power subsystem. NASA (1972) https://ntrs.nasa.gov/api/citations/19720025198/downloads/19720025198.pdf
22. A. Petrocelli, R.A. Bettinger, N.C. Kutz, C. Hartsfield, Atlas of historical and proposed nuclear devices and power systems in the earth-moon system and wider solar system. Prog. Nucl. Energy **164**, 104857. https://www.sciencedirect.com/science/article/pii/S0149197023002925 (2023). https://doi.org/10.1016/j.pnucene.2023.104857
23. NuclearNewswire, in *Westinghouse's Lunar Microreactor Concept Gets a Contract for Continued R&D*. https://www.ans.org/news/2025-01-15/article-6686/westinghouses-lunar-microreactor-concept-gets-a-contract-for-continued-rd/. Accessed Date 16 Mar 2025
24. The Space Review, in *Repurposing Nuclear Reactors Used in Space Propulsion for High-Density Power on the Moon and Mars*. https://www.thespacereview.com/article/4888/1. Accessed Date 16 Mar 2025
25. X Energy, in *Intuitive Machines and X-energy-led Team Awarded $5 Million to Provide a Solution to Deliver Fission Surface Power to the Moon by 2028*. https://x-energy.com/media/news-releases/intuitive-machines-and-x-energy-led-team-awarded-5-million-to-provide-a-solution-to-deliver-fission-surface-power-to-the-moon-by-2028#:~:text=%E2%80%93%20June%2021%2C%202022%20%E2%80%93%20The,to%20the%20Moon%20by%202028. Accessed Date 16 Mar 2025
26. Space.com, in *Russia and China Announce Plan to Build Shared Nuclear Reactor on the Moon by 2035, 'Without Humans'*. https://www.space.com/russia-china-shared-nuclear-reactor-2035-moon. Accessed Date 16 Mar 2025
27. Rolls Royce, in *Micro-reactor*. https://www.rolls-royce.com/innovation/novel-nuclear/micro-reactor.aspx Accessed Date 16 Mar 2025
28. US and Turks and Caicos to inquire into failed SpaceX launch leading to debris. The Guardian 18 Jan 2025. https://www.theguardian.com/science/2025/jan/17/turks-caicos-starship-spacex
29. M. Lenkart, S. Biller, Space debris material sourcing for in-space manufacturing: a quantitative evaluation framework. Front. Space Technol. **6**, 1690460 (2026). https://doi.org/10.3389/frspt.2025.1690460
30. G. Polansky, M. Houts, *A Preliminary Investigation of the Topaz II Reactor as a Lunar Surface Power Supply*. (1995). https://www.researchgate.net/publication/241908334
31. A. Barco, R.M. Ambrosi, H.R. Williams, K. Stephenson, Radioisotope power systems in space missions: overview of the safety aspects and recommendations for the European safety case. J. Space Saf. Eng. **7**(2), 137–149. https://www.sciencedirect.com/science/article/pii/S2468896720300288 (2020). https://doi.org/10.1016/j.jsse.2020.03.001
32. NASA Scientific Visualisation Studio, in *Moon Essentials: Seasons*. https://svs.gsfc.nasa.gov/5229/. Accessed Date 28 Mar 2025
33. C.P. Spedding, W.J. Nuttall, S. Lim, Energy requirements of a thermally processed ISRU radiation shield for a lunar habitat. Adv. Space Res. **65**(11), 2467–2474.https://www.sciencedirect.com/science/article/pii/S0273117720301708 (2020). https://doi.org/10.1016/j.asr.2020.03.015
34. European Space Agency, Lunar Solar Power Satellite [Internet]. Paris: ESA; 2023 Jul 14. Available from: https://www.esa.int/ESA_Multimedia/Images/2023/07/Lunar_solar_power_satellite

35. Landis GA. Laser Power Beaming for Applications on the Moon [Internet]. Paper presented at: Space Power Workshop; 2025 Apr 29-May 1; Torrance, CA. Cleveland (OH): NASA Glenn Research Center; 2025. Available from: https://ntrs.nasa.gov/api/citations/20250003925/downloads/Laser%20Power%20beaming-SPW-2025.pdf

Chapter 2
Geopolitical and Economic Drivers of Fusion on the Moon

Contents

2.1 The Role of Prestige and Past Commitments

The impetus for ambitious space exploration programmes is rarely singular. While scientific discovery, resource acquisition and technological advancement often feature prominently in official justifications, the intertwined forces of national prestige and the weight of past commitments frequently exert a significant, if sometimes less explicitly stated, influence. The enduring human fascination with the Moon, coupled with the historical context of the space race, has created a powerful narrative that continues to shape the rationale for lunar missions, including the contemporary Artemis programme. Understanding the role of prestige and past commitments is crucial to comprehending the complex motivations driving humanity's return to Earth's celestial neighbour.

W. J. Nuttall et al., *Fusion on the Moon*,
SpringerBriefs in Applied Sciences and Technology,
https://doi.org/10.1007/978-3-032-38217-7_2

The pursuit of national prestige has been a consistent driver of space exploration since its inception. The launch of Sputnik in 1957 ignited the space race, not merely as a technological competition, but as a symbolic battleground for ideological supremacy. The United States' subsequent 1960s commitment to landing a man on the Moon before the end of the decade was a declaration of its perceived technological superiority and national will. The Apollo programme, culminating in the 1969 lunar landing, became a defining moment of the twentieth century, cementing American prestige on the global stage. This historical precedent continues to resonate in contemporary discussions surrounding lunar missions. The notion of reclaiming that pioneering spirit, and once again demonstrating technological leadership and inspiring national pride, remains a potent, if often unspoken, element in the justification for returning to the Moon—it's only the perceived competitor that has changed.

The language employed by policymakers and space agencies frequently reflects this underlying desire for prestige. Terms like 'leadership', 'innovation' and 'historic' are interwoven into official pronouncements, invoking the legacy of Apollo and framing current lunar ambitions as a continuation of that grand narrative. Official remarks are also often specifically designed to call back to the Apollo programme, gently coded in modern lexicons, such as '... land the first woman and the next man on the Moon'. It must be acknowledged, however, that in 2025 the Trump administration has rolled back the gender and ethnic diversity rhetoric around Artemis, although the programme retains its name—recalling the name of Apollo's sister in classical mythology.

It remains the case that in an era of increasing global competition, particularly in technological spheres, demonstrating dominance in space remains a significant motivator. A successful and sustained lunar presence could be interpreted as a symbol of national strength and a testament to a nation's capacity for complex, long-term endeavours. This is not simply about flag planting; for the US, it is about projecting an image of competence, ambition and demonstrating the sort of long-term, sustained effort that transcends administrations. It is also an approach that is becoming increasingly visible in Chinese lunar strategy.

The concept of prestige extends beyond purely nationalistic aspirations, however. In an increasingly multipolar world, collaborative lunar missions can also serve as a means of enhancing international standing and fostering diplomatic relationships.[1] Partnering with other nations on ambitious projects like lunar bases allows participating countries to share in the reflected prestige and strengthens alliances through shared goals and achievements. The Artemis Accords, while controversial in some quarters, can also be viewed through this lens—an attempt to establish norms of behaviour in space that position the initiating nations as leaders in shaping the future of space exploration.

Even were you to disregard the national posturing, there is a sober economic reality for many American policymakers—significant investments have been made over decades in developing the infrastructure, knowledge and expertise necessary

[1] See the discussion of the Axiom Space Ax-4 mission in Chap. 1.

for lunar missions. Agencies like NASA have cultivated a deep institutional memory and a workforce skilled in the intricacies of spaceflight and planetary science. Abandoning lunar ambitions entirely could represent a potential waste of these accumulated resources and expertise and put jobs at risk across the entirety of the US. That said, the recent emergence of highly competent private sector players in the space arena reveals that potentially NASA now has competitors and collaborators that do not represent the interests of a foreign power, but rather the interests of the American private sector. Historically, past investments have created a form of path dependency, making continued lunar exploration a more logical and economically justifiable direction than starting entirely anew with a different celestial target or with a different institutional architecture. As these issues shape decision-making in space, very similar dynamics are playing out in fusion energy technology policy and strategy back on Earth.

Historically, the political dimension of past commitments has been crucial. Presidential pronouncements and legislative mandates created expectations and established long-term goals that subsequent administrations found difficult to overturn entirely. The 'Moon to Mars' strategy, for instance, has been a recurring theme in US space policy, creating a sense of momentum towards lunar return as a necessary precursor to Martian ambitions. Cancelling or significantly altering such long-standing commitments can be politically damaging, perceived as a lack of resolve or a failure to follow through on established national objectives, but with that said President Trump appears comfortable with bold and disruptive changes.

While scientific, economic and technological rationales are undoubtedly important drivers of lunar missions, the influence of national prestige and the weight of past commitments cannot be understated. The historical legacy of the space race continues to shape the desire for recognition and leadership in space, while the achievements and investments of past lunar programmes create a powerful momentum for continued exploration. These factors, often operating beneath the surface of official discourse, contribute significantly to the enduring pull of the Moon, ensuring that it remains a key destination in humanity's ongoing journey into the cosmos. Understanding these motivations provides a more complete picture of the complex forces driving the return to the lunar surface.

Text Box 3: Lunar Governance—Lessons from Antarctica

The Moon and Antarctica are examples, in a political and governance sense, of international spaces. Antarctica is a relevant point of reference as it is the most tightly governed of the international spaces (others international spaces include the high seas, the seabed and outer space generally). It is through the Outer Space Treaty that lunar governance is shaped today.

Antarctica in the First Half of the Twentieth Century

It is easy to forget that Antarctica was once a territory open for exploitation and settlement. The human history of Antarctica is really a twentieth, and now twenty-first, century story. The first documented visit to the mainland of Antarctica was in 1895 when a Scandinavian whaling ship, the Antarctic, anchored at Cape Adare. The early twentieth century saw a flurry of expeditions, including the famous 1912 race between Amundsen and Scott to the pole. After the First World War, American interest grew. Prominent private expeditions were funded by figures such as William Randolph Hearst and military interest also grew, led by US Navy Rear Admiral Richard Evelyn Byrd. In the lead up to the Second World War, territorial claims started to cause friction with Nazi Germany acting particularly aggressively and coming into contention with Norway. In terms of the opening up of the Moon, are we today perhaps where Antarctica was at immediately after the First World War?

Antarctica and Science Diplomacy

After the Second World War, territorial claims intensified, but moves were also afoot to establish a new international order for the continent. Under the auspices of the 1958 International Geophysical Year, U.S. president, Dwight Eisenhower led international efforts to end the rivalry. In 1959 nine countries with declared interests in the region, including the USSR, concluded the Antarctic Treaty. Coming into force in 1961, it stands as an example of science diplomacy in opposition to militarism. The Antarctic Treaty is remembered as the first arms control treaty of the Cold War era.

Paul Berkman has suggested that, such was Eisenhower's enthusiasm for a global agreement on the Antarctic Treaty, he held US technologists back in moves to explore another international space—outer space. Berkman alleges that the US President delayed the 1956 launch of an American satellite on a Jupiter-C rocket, only to see the lead in space taken by the USSR with the launch of Sputnik in 1957 [1].

We might also speculate that the US was willing to see the demilitarization of Antarctica in recognition of other practicalities that might also have been acutely visible to the USSR. One must note that in 1959, the United States established a nuclear missile base in Greenland, at Camp Century as part of Project Iceworm [2]. Greenland is almost as hostile a geography as Antarctica. While our focus in this book avoids direct consideration of military activities on the Moon one can see parallels between the Moon and Antarctica regarding a scenario of geopolitical rivalry and competing claims versus a scenario resting upon science for global good.

Antarctica—A Zone of Peace in Times of Conflict

The geopolitics of the mid-1950s started with the emergence of the United Nations as a tool of conflict avoidance with the UN Police Action in Korea, commonly known as the Korean War, being a salutatory lesson for all. The communist regimes of China and the USSR fought to a stalemate against a UN mandated force led by the United States.

The Korean War was exhausting, but nothing compared to the prospect of a war in Antarctica. The terrain, climate and months of darkness rendered the prospective battle space impossible for armies in the mid-twentieth century. The only realistic prospect for viable combat would be to use Air Force delivered nuclear weapons against enemy footholds and strongpoints. The detonation of nuclear weapons is specifically banned by the original text of the 1959 Treaty. Similarly, the disposal of radioactive waste is also proscribed. The Treaty regime later placed wider obligations on signatories to protect the environment of Antarctica prompting parallels for lunar governance. Arguably the science-based peaceful way ahead was found for Antarctica because more militaristic scenarios were not viable for the technology of the time. Perhaps a similar truth applies to the Moon today. Such ideas reinforce our decision not to focus on military thinking and its potential role in lunar settlement and development, but we should ask, can the Moon and Antarctica remain zones of peace?

Antarctica—Will it remain a zone of peace?

The changing geopolitics of the 2020s and the changes in military capabilities (robotics, drones, integrated weapons systems and improved infantry capability) open up a real risk for the future that, we posit, has not been adequately discussed. Under the terms of the Antarctic Treaty from 2048 onwards any state party to the treaty can call for a review conference of the environmental protocol [3]. The treaty itself contains no expiration date or requirement for review. But states can withdraw from the treaty and, as a consequence of considerations around treaty protocols, 2048 will likely be a stress point for the wider treaty regime. It is noteworthy that the United States has withdrawn unilaterally from arms control treaties or has allowed them to expire.

Treaties from which the US Exited

- Anti-Ballistic Missile Treaty (ABM)—unilateral exit on the basis of the threat of ICBM attacks from North Korea, the expected salvos of which were small enough not to saturate ballistic missile defence systems.
- Intermediate-Range Nuclear Forces Treaty (INF)—technically a unilateral exit, after unsuccessful negotiations with the Russian Federation on their development of weapons systems that the US felt were in breach of the Treaty.

Expired Treaties

- Strategic Arms Reduction Treaty (START-1).
- New START—continuation of START. Russian Federation 'suspended' participation in the Treaty as diplomatic retaliation for US support of Ukraine in 2023, but stated it would continue to abide by warhead limitations. It expired in February 2026.

For further information see: [4].

Of course, the threat to the enduring power of the Antarctic Treaty does not just come from the United States. The Russian Federation could be regarded as a greater obstacle to continuity in the region. The point here is, however, that it is not manifest that Antarctica's status as an inaccessible zone of peace and science will endure forever. It is the use of diplomacy and international law that has protected Antarctica for nearly 100 years. The Antarctica Treaty itself was the product of singular focus of the relevant Premiers at the time, and that no such language or sentiment seems to be present when it comes to the Moon today.

One vital aspect of the Antarctic Treaty has been its role in preserving a sensitive environment. All activities on the continent are closely regulated, and tourism, for example, is in effect heavily restricted and subject to strict environmental standards. Even after the Antarctic Treaty came into force remote bases struggled to manage their wastes, as illustrated in Fig. 2.1. In recent decades concern for the Antarctic ecosystem has continued to grow.

Fig. 2.1 Old fuel drums and food tins at Fossil Bluff waste dump before removal from Antarctica in the summer of 1994/5. Photo credit: Pete Bucktrout, British Antarctic Survey

In Chap. 1, Sect. 1.6 we argued that concerns for environmental protection and clean-up could limit the deployment of nuclear fission power on the Moon. Fission is a key competitor to Fusion for MW power on the Moon, as discussed in Chap. 1.

2.2 Towards Business Models for the Moon

2.2.1 *Government or Private Sector Leadership?*

In this section, we build upon ideas developed earlier, in Chap. 1. The extreme environment of space, combined with the substantial upfront investments required, makes international collaboration and public–private partnerships part of an essential drive to reduce costs and to diversify risks, particularly in initial periods of lunar economy development. The first phases of the lunar economy will likely be heavily reliant on governmental funding, even if executed by commercial entities. However, space agencies are increasingly undertaking commercial ventures and as a consequence of that, one sees a reallocation of funding toward more cost-effective, private sector solutions.

A significant portion of NASA's budget is allocated to the Artemis programme (see Text Box 1 in Chap. 1), which aims to return humans to the Moon and establish a sustainable presence there. As of 2025, the total expenditure from 2012 through 2025 was projected to reach approximately $93 billion [5]. A significant portion of that funding is, or has been, allocated to partnerships with commercial companies.

To further reduce and diversify its cost base, NASA shifted its approach from contracting rockets to purchasing rides from private companies such as SpaceX and Blue Origin. As part of NASA's Artemis ambitions for the Moon, the agency has created the Commercial Lunar Payload Services (CLPS) programme [6]. This has partnered with more than a dozen American companies and more than $2.6 billion of contracts have been issued covering the period up to 2028. This operational transition towards partnership with the private sector has led to significant cost reductions. SpaceX's reusable Falcon 9 rocket, supported by billions in NASA funding, has revolutionised access to space, making satellite deployment and space missions far more cost-effective. It is interesting to note that Falcon 9 rockets flew roughly the same number of times just in 2024, as all the missions of the long US Space Shuttle programme. Many start-up companies are now competing to place satellites in low Earth orbit. Via the development of heavy lift systems SpaceX is now aiming to bring a similar transformation to lunar transportation, further reducing costs and enabling economically viable business models in space [7]. Reference [7] also suggests that the lunar economy over the next 10 years might reach $137 billion, with 400 missions expected. There is clearly enormous interest in the Moon. SpaceX founder Elon Musk has recently switched his emphasis from Mars to the Moon [8] and his economic, and US political, influence is substantial.

2.2.2 *Space Tourism*

As introduced in Chap. 1, in this book we do not imagine space tourism as a major early driver of fusion on the Moon; rather we see the prospect of large-scale lunar tourism emerging later in our window of interest perhaps boosted by success in deploying MW energy systems on the Moon in the intervening period. We believe that economically meaningful space tourism to the Moon will emerge in the long-term, unless blocked by regulation in ways reminiscent of legal and practical restrictions on tourism in Antarctica.

We suggest that in the early part of our period of interest (2040–2055) lunar tourism does not have the potential to get ahead of the other economic drivers explored in this book. We suggest that while space tourism may contribute to the lunar economy, sectors like transportation and resource extraction are likely to be more important in the early phase. In Chap. 5, we stress our interest in lunar mining in challenging polar PSR. We suggest that cold, dark PSRs are not natural lunar tourism destinations. Tourists will be much more interested to see the Apollo 11 landing site in the Sea of Tranquillity, for example. It is also possible that short duration tourist visits will want to use the somewhat easier proposition of near-equatorial landing zones.

Our sense that large-scale lunar tourism will lag renewed exploration and industrial activity does not diminish our sense that space tourism will be a twenty-first century phenomenon. As noted earlier, Dennis Tito was the first commercial astronaut in 2001 riding aboard a Russian Soyuz module. The difficult financial situation facing the Russian space programme after the collapse of the Soviet Union had led the Russian space community to seek any and all revenues—including space tourists. At the time, the Americans expressed gentle disapproval, but it was not long before the west caught up.

As noted in Chap. 1, term space tourist is perhaps not helpful to understanding and indeed is arguably somewhat pejorative. ‘Commercial Astronauts’ is a more generic term more frequently used by the space community. Their numbers are so far very small, and they are typically extremely rich individuals seeking the experience of space flight. They see themselves as adventurous and the experience is not without significant danger.

Earlier we posited the relevance of Antarctica to thinking about lunar futures—it represents a vision of a terrestrial utopia acting as a model for the Moon. There is some Antarctic tourism, but it is very carefully controlled and managed. In contrast, Mount Everest has become a popular destination for climbers, leading to overcrowding and environmental issues like increased waste and pollution [9]. Of course, Mount Everest is not an international space like the Moon or Antarctica, but it is a challenging environment attracting international interest, and increasingly international commercial interest. The commercialisation of Mount Everest has led to significant environmental degradation.[2] Is this a warning tale for lunar exploration? The Moon’s unique and extremely demanding environment offers opportunities for exploration and scientific research, but it also sparks imagination and inspires bold ambitions. Unregulated commercial activities, such as mining or tourism, could damage a unique environment and make scientific research even more challenging. To prevent the Moon from

---

[2] In February 2026, a bill was introduced in the Nepalese National Assembly to restrict such activities [10].

facing challenges similar to those seen on Everest, or worse, it will be essential to promote sustainable technologies and solutions and establish international regulations that balance commercial interests with the preservation of the lunar environment (e.g. strict waste management).

The first offerings in lunar tourism are starting to emerge, but are they credible?

We note that an American start-up GRU Space captured headlines by announcing that it is open for taking bookings at its planned hotel on the Moon for the 2030s [11]. The company has also issued a White Paper describing the plans. They share our sense that there is a parallel between lunar tourism and today's Everest tourism.

GRU Space's plan is to use high-end lunar tourism as the first commercial wedge into a wider lunar economy. In its white paper, the company argues that by building a first Moon hotel, it can generate private demand and revenue while proving the same capabilities needed for a broader lunar industrial base, especially pressurised habitation, in situ resource utilisation (ISRU), and regolith-based construction. Its roadmap starts with a small lunar demo in 2029, a larger ISRU and inflatable-habitat test near a lunar pit in 2031, and then a first hotel in 2032. Their vision includes an Earth-built inflatable habitat for 4 guests staying for several days. GRU Space then wants to expand to a larger v2 hotel with more shielding and longer stays by using robotic construction and lunar materials, before eventually growing from one-off tourist missions into a permanent Moon base and, later, a similar model on Mars. This paper presents the hotel not just as a tourism product, but as the first economically rational module of a future lunar base, blending private revenue with public sector demand. We notice that the economics and timelines depend on future launch prices, lunar transport availability, regulatory progress and technical execution and that they are assumed rather than demonstrated. However, we find it interesting and relevant to our discussion of potential energy demand on the Moon.

On the one hand, GRU Space's proposal suggests that some lunar tourism may be possible without megawatt scale power, because the company explicitly rejects energy-intensive construction methods such as sintering and melting regolith in the early phase and instead proposes geopolymers as an intermediate, lower-energy approach. On the other hand, they also implicitly support the wider argument that a larger and more mature lunar economy will eventually need much more power. GRU Space notes that sintering and melting regolith would require megawatt scale power on the Moon, with extensive solar arrays and radiators. According to the authors, it would only transition to such methods later, depending on available power. Similarly to our thinking, this paper suggests that early lunar tourism does not require fusion-level power, but it also supports the view that large-scale expansion, including durable construction, and a true lunar industrial base, would drive much higher energy demand over time.

2.2.3 The Role of Branding

In recent decades, space tourism has moved from science fiction to a commercial reality, with private companies competing to make space more accessible to civilians. The British entrepreneur Richard Branson has done much to bring down the cost of space tourism with sub-orbital flights on Virgin Galactic's space plane (see Fig. 2.2). However, going to space is still only for the very rich. A seat on Virgin Galactic's Delta flight costs $600,000 [12]. Virgin Galactic also presents a fascinating case study in branding which is powerful in building trust and visibility but also reveals the limits of commercial influence in large-scale space infrastructure projects. Branson and his Virgin Group are usually, but not always, masterful in their use of branding. Virgin Trains, however, was a difficult experience for the Virgin group as the train service struggled with service quality and customer satisfaction. This is a good reminder that branding can't always compensate for technical and operational challenges. Branson has made branding a key part of his space tourism business. With Virgin Galactic, Branson leveraged the emotional appeal and futuristic vision of spaceflight to extend his brand identity into an aspirational and adventurous new market. It built trust for customers, and it built the brand for his numerous other businesses. The Virgin brand is proudly visible on the Virgin Galactic replica (see Fig. 2.2) as it is on the actual spacecraft. It is reinforcing the connection between space exploration and Virgin's brand values of innovation, excitement and personal empowerment.

Fig. 2.2 Full-scale replica of Virgin Galactic SS2 on display at Farnborough International Airshow, Hampshire, England, 15 July 2012. *Image source* Steve Mann/Shutterstock.com

Such a branding strategy is effective in building customer trust, particularly in an industry where safety, credibility and prestige are crucial. The true value to the Virgin Group was that these values would boost the brand for other group businesses. The familiar branding helped to normalise the idea of civilian spaceflight and aligned it with Virgin's existing reputation for breaking new ground in consumer experience. However, while the Virgin Galactic brand has generated considerable attention and presumably has helped drive bookings, the company has faced multiple delays, technical setbacks and criticism regarding its long-term viability. The strength of a brand is important to draw public attention and funding, but it cannot alone overcome the immense technical and financial challenges of developing sustainable spaceflight systems. As with space tourism, we suggest that while space as a terrestrial brand builder may be a genuinely commercial revenue builder, it is not sufficient, in isolation, to justify advanced infrastructures such as a tokamak on the Moon. We here return to our sense that space tourism will be a component of the lunar economy, but its overall impact is expected to be part of a broader economic framework. Sectors such as transportation and resource extraction are expected to be major drivers of lunar economic activity [13].

Space is a sector with active brands—indeed NASA has a brand. The brand is developed and promulgated across a wide range of merchandise including clothing. Virgin, however, is not a space brand per-se. It is an established brand in a range of terrestrial sectors. While the space activities arguably do much for the Virgin brand, boosting societal goodwill in support of Virgin group's range of diverse direct-to-consumer commercial products, we suggest a marketing boost to terrestrial businesses is not itself sufficient to fund lunar settlement and industrial activity.

2.2.4 The Long Term

Commercial business models oriented directly towards individual consumers are unlikely to provide sufficient economic incentive to drive lunar expansion on the scale that would require lunar-based fusion. However, we do expect that private commercial activity will have a deep and enduring role to play in lunar settlement.

We posit that, in the long term, the experiences of consumers on Earth will involve a range of complex commercial links to lunar activity. Today, we already engage with the space economy via our smartphones (GPS etc.) in time we will engage with the benefits of lunar activity in terms of energy provision (e.g. helium-3 fusion) or advanced materials (e.g. rare metals once known as rare Earth metals).

In time, and as lunar settlement expands, eventually the settlers themselves will become consumers driving economic progress. Such a scenario was anticipated by the Jet Propulsion Laboratory in California in one of its *Visions of the Future* space posters [14]—see Fig. 2.3.

Fig. 2.3 One of a series of futuristic posters created by a team of visual strategists at Jet Propulsion Laboratory, known as 'The Studio'. This image imagines selling vacations on Earth to lunar settlers. Image courtesy of NASA/JPL-Caltech

2.3 Commercial Scenarios for Sustainable Lunar Settlement

One key consideration for sustainable lunar settlement relates to the health of the explorers, then the settlers and later still the mass market tourists. The lunar surface is an extremely hostile environment quite apart from the need to maintain life support systems. The absence of a lunar magnetic field and atmosphere means that cosmic and solar radiation is much more hazardous than it is on Earth. Another consideration is meteoroid and asteroid impact including very small objects. The absence of an atmosphere ensures that they reach the lunar surface intact and at very high speed. Because of such risks, humans on the Moon for any extended period (beyond hours) should ensure that they are suitably shielded [15]. In this book, we emphasise one solution—to go underground.

Another health consideration is very important, the low gravity on the Moon. This will affect all settlers and there is nothing that can be done about it beyond slowing the rate of physiological damage through exercise plans and specialised nutrition. The proximity of the Earth to the Moon allows for a system model based on crew rotation. We posit that, in the coming decades at least, young workers will all rotate every few months. There is another parallel with Antarctica here. There will, however, be exceptions that we shall consider later (Sect. 3.4).

The considerations of human health are part of our motivation to use the word Semi-Autonomous when describing lunar futures. The other key consideration motivating that choice is relative proximity. We suggest that the Mars scenario will inevitably be very different. There is a far greater likelihood that travellers to Mars will be asked to accept the probability of a one-way journey, where a return to Earth cannot be assumed.

Another key aspect of a sustainable lunar future is whether the lunar economy can evolve into a semi-autonomous commercial ecosystem. What will this reality look like? Will it follow a 'first come, first served' model, where early pioneers gain a dominant advantage? If so, the benefits for first movers could be immense, establishing long-term economic leadership in space.

Evaluating the economic viability of fusion energy generation on the Moon involves several complex factors, including technological maturity, infrastructure costs, operational expenses and potential revenues. Nuclear fusion remains in the experimental stage on Earth, with significant scientific and engineering challenges yet to be overcome. Projects like Iter (or ITER—originally the International Thermonuclear Experimental Reactor) have experienced escalating costs. ITER is a research machine—it will not generate any electricity, but the ITER journey towards commercial fusion prompts estimates reaching $20–25 billion per gigawatt (GW) of capacity [16]. Private companies aim to reduce project costs to $2–5 billion for 100–500 megawatt (MW) plants, but these figures remain speculative [16].

Achieving and maintaining the extreme conditions necessary for fusion reactions presents substantial technical challenges which can potentially lead to higher operational costs compared to other solutions. While fusion offers the prospect of a nearly limitless and cleaner energy source, the current lack of a functional fusion reactor makes it a long-term prospect. In the near to medium term, fission appears to be economically more feasible option for lunar energy production due to its established technology and shorter deployment timeline (see Sect. 1.6). The high initial investment and uncertain timeline for fusion operational readiness pose significant economic risks and make it hard to become a path for VC funding.

On the other side, the landscape around fusion energy research and development (R&D) has evolved significantly, especially in the United States. There are indications of fusion's increasing technical readiness.

While some private companies, like Helion Energy, aim to have operational fusion plants by 2028, these timelines are ambitious and might require further technological breakthroughs. The U.S. Department of Energy's 'Bold Decadal Vision' aspires to create a commercially relevant, private sector-led fusion pilot plant in the 2030s. Over $6 billion has been invested in private fusion companies, with 80% of funding concentrated in the United States, highlighting fusion's commercial potential. While the U.S. leads in private sector fusion R&D, global interest is rising, with significant investments in Canada, the U.K., Japan, the E.U., China and beyond [17].

Private–public partnerships are recognised to be a key enabler of the American fusion strategy. In 2024, it was observed that the benefits of such partnerships include 'better leveraging by the private sector of the expertise and capabilities in the public sector; greater available financial resources to achieve objectives and accelerated timelines; alignment of stakeholder incentives by sharing costs; and R&D being pursued in a way that is relevant and valuable for commercialisation' [17].

We have earlier rejected certain commercial activities on the grounds that they do not appear to be sufficient, in isolation, to drive the deployment of fusion on the Moon. For example, we suggest that while lunar tourism could be very important, it is more likely to follow from success with the industrial development of the Moon than to be a driver of such development. Small-scale lunar hotels with approximately 10 guests (as envisaged by GRU Space) could be developed without advances in energy, water and other infrastructures, but larger-scale tourism (such as we already see today on Mount Everest) would require more advanced infrastructures.

We shall now present four propositions that we suggest could be of sufficient enduring strength to drive investment and activity, perhaps even from the start of our period of interest (2040) and furthermore at a scale required to make fusion on the Moon viable and sustainable.

2.3.1 Proposition 1: Helium-3 Mining and Export to Earth

We posit that this proposition has the greatest potential to drive lunar activity to a scale at which fusion on the Moon would be considered as preferred energy source. In addition, the synergies with possible in situ use of ^{3}He on the Moon renders the proposition even more compelling.

Helium-3 is a rare form of helium with two protons, but only one neutron in its nucleus. The three heavy nuclear particles give it its name. The more abundant form of helium stable ^{4}He has two neutrons. ^{3}He is found on the Moon for fundamentally two reasons both related to the irradiation of the lunar surface by high energy particles from outer space. On Earth such particles generally do not reach the surface owing to the planet's magnetic field and atmosphere. On the Moon however the incoming particle can split the nucleus of an atom on the lunar surface (this is known as spallation based on the geological term—'to spall', meaning to chip). The small nuclear fragments are likely to be unstable and they will therefore decay through radioactivity. Some fraction of these small atoms will settle as stable ^{3}He. The other pathway by which ^{3}He is found on the Moon is direct radiation of high energy ^{3}He atoms or ions from outer space.

We note, in addition, that lunar water is likely to be a particularly rich source of ^{3}He if conditions are such that it has received a neutron flux caused by cosmic radiation over aeons. While water must be in dark PSR, these regions can still perhaps be reached by deep space cosmic rays.

As regards an export-based lunar economy, looking out to 2070, we posit that only ^{3}He will have a price per gram sufficient to sustain a lunar export business model within our timeframe of interest. The business model is straightforward commodity extraction and sale. Any industrial enterprise interested in such a business will need to:

- invest in lunar mining infrastructure,
- produce ^{3}He gas,
- ship it to Earth,
- sell it to whomever operates fusion power plants or other ^{3}He-consuming industries on Earth.

In this proposition, the Earth economy funds the extraction of ^{3}He on the Moon. As such, the lunar activities form part of an industry based in the terrestrial economy with revenues earned on Earth. While we accept the orthodox logic that D–T fusion will appear on Earth before alternative ^{3}He based fuel cycles we do not exclude the possibility that the emergence of ^{3}He imports from the Moon in the late twenty-first century could not only fuel stable ^{3}He-based fusion, but it might also improve the economic prospects of D–T fusion. This is because of the relative ease by which ^{3}He can be converted to tritium, without the need for fuel breeding in fusion reactors. We suggest that, barring a breakthrough, affordable tritium supply will always be a concern for terrestrial D–T fusion. The arrival of ^{3}He from the Moon for terrestrial processing to tritium could be such a breakthrough.

The lunar ^{3}He for export value chain involves several stages. Each stage could be handled by the same company or by different companies in a vertically integrated chain. For instance, one can envision a mining company that sells raw ^{3}He to an energy company, or a fully integrated enterprise that handles everything from digging moon dust to delivering fuel to utility customers on Earth.[3]

The potential profitability of ^{3}He mining is very high, but the proposition is only possible if technical feasibility and market readiness align. If a lunar mining operation can ultimately scale to produce a tonne per year, then the gross revenue would be enormous by any mining industry standards. A single metric tonne (1000 kg) of ^{3}He could be valued at around $1–2 billion in future energy markets [18]. However, these valuations are speculative and depend on the successful development of helium-3-based fusion energy. This scenario has the potential to become highly profitable in the long term, but the path to get there is still challenging. This proposition will attract concern as to its costs even if things go well, but of even greater concern to investors will be the economic risks, as there is little opportunity for risk mitigation or avoidance in such a challenging location and with such an unusual business proposition. This economic risk will, however, be resolved in time through experience and innovation. At this stage, and concerning the mining for lunar export business model, we recommend that, as a first step to understanding the possible economic revenue risks on the demand side that anticipated ^{3}He prices on a 2040–2070 timescale are further researched. This is as important as considering the supply side risks with which cost-focussed engineers are usually more comfortable.

While we have posited ^{3}He production measured in kg, or even tonnes, companies like Interlune are targeting much smaller quantities to serve existing terrestrial markets (e.g. quantum tech and medical applications) which provide a revenue 'bridge' until fusion reactors come online.

Meanwhile, according to one report, the first trip to the Moon to establish mining infrastructure could take place as soon as 2027 [19]. It is important to remember that ^{3}He is highly valuable even in the absence of a terrestrial fusion economy, but the scale of this industry will always be smaller than energy-related demand. By supplying ^{3}He at approximately $20 million per kg to specialised customers, a mining venture can establish early cash flow, while refining its extraction technology and all at a small scale. As fusion energy demand emerges, the lunar operation can then ramp up extraction. In this way, the expansion of the business can be revenue-based, and the industry can grow organically. In summary, the first reason to mine ^{3}He on the Moon is to extract it for use in Earth's existing industries; the second reason to mine it is for fusion energy use terrestrially, on the Moon, or both.

We remind the reader that in our opinion, there are fundamentally three resources of interest on the Moon which could be an economic driver of settlement:

- *Water-ice.*
- *Helium-3.*
- *Metal deposits in craters.*

These resources will become economically relevant on different timescales, but we posit that the synergies could be important.

[3] Such scenarios have been presented in science fiction movies (such as *Moon* directed by Duncan Jones and released in 2009) and TV series (such as series 3 of *For All Mankind* from Apple TV+).

The proposition outlined here could eventually lead to very large profits for investors by supplying Earth's enormous energy market. However, this proposition faces major technical and political challenges. For its fullest implementation, it depends on having working fusion reactors to use the fuel and the ability to mine the Moon safely and affordably. It further assumes that future laws and agreements (or the absence of such things) will allow companies, or governments, to own and sell materials taken from the Moon. Such scenarios were discussed at length in a webinar entitled *Who Owns the Moon: The Legal Reality of Mining Helium* held on 18 March 2026 and organised by Lunar Helium Mining LLC (LH3M). We summarise our impression of a consensus from that meeting that while no claims of territorial sovereignty can be made, there is widespread acceptance that extracted resources can be owned and traded.

2.3.2 Proposition 2: Fusion Power for an Enduring Lunar Settlement

In this scenario, nuclear fusion reactors are deployed on the Moon to supply power for lunar bases and secondary industrial activities. Profitability in this settlement power model might be limited in the early decades because the customer base on the Moon is small. In addition, the financial input is essentially entirely from Earth, and this is mostly governmental, either directly or through contracted private actors. We posit that this proposition is smaller and has less dynamic growth than Proposition 1. In addition, it will be later in the timeline.

In this proposition, the initial fusion power plant on the Moon would likely be a publicly funded demonstrator (e.g. a NASA or ESA-funded reactor providing energy for an international Moon village) or perhaps it would arise via public–private partnership, noting U.S. strategy as evidenced by the CLPS programme discussed earlier (Sect. 2.1).

We posit that the bulk of early-stage funding of lunar activity will not be market-based; it will essentially be governmental budget allocation. As such, the economics of the proposition must be calculated in terms of cost savings. For example, a fusion reactor that reduces the need to transport many tonnes of chemical fuel or batteries from the Earth provides an economic benefit by avoiding the associated upmass costs. In this proposition, the focus is not on profitability but on enabling other high-value activities (science, exploration, prestige), and government support or ownership is critical.

As the lunar presence grows, a more commercial arrangement, even a market, could emerge. For example, private operators might run power stations and charge users per kilowatt-hour, perhaps under long-term contracts as would be required to compensate for the significant capital at risk. Such a lunar presence takes us towards the prospect of citadel settlements assembled around a large energy source as might be provided by fusion on the Moon.

2.3.3 *Proposition 3: Lunar Power Distribution and Energy Services*

A third proposition involves using the Moon as a site for power generation and then distributing that energy either across the lunar surface or even transmitting it to Earth, to Mars or to a spacecraft. Such thinking takes us to Microwave Wireless Power Transmission (MWPT) technology. While such technology has never been demonstrated under the challenging conditions of the lunar surface, it is known to be possible, in principle at least.

Of the propositions presented thus far this is perhaps the most speculative, but it is worth examining as a complementary model to the other three. It has the potential for powerful synergies with the idea of fusion on the Moon. The use of MWPT in space is facilitated by the lack of scatter and attenuation in the ultra-high vacuum of space.

This model for a Lunar Energy Economy would export energy via microwave beams sourced from large lunar antennas. As such, we are here imaging the inverse of the more usual proposition of beaming power from the Earth to the Moon. That inverse model might have more to do with Proposition 2.

We wonder whether the ^{3}He and MWPT propositions are exclusive? Our initial conclusion is that they are probably not. The MWPT export proposition is arguably even riskier than fusion on the Moon. We posit that the infrastructures of large-scale Lunar MWPT would be more fragile, more exposed and less easily defended than the citadel model of a possible Fusion on the Moon installation.[4]

Importantly, MWPT export might not be direct to the Earth, but rather to spacecraft, satellites (e.g. in lunar orbit), and larger space stations (e.g. Lunar Gateway facilities). We suggest that those interested in space applications of MWPT compare their ideas to the powering of deep space using space-based solar power (in the short term at least). Such assessments lie outside the scope of our interests, beyond simply our desire to ask whether such developments would enhance or militate against the success of Fusion on the Moon. The relationship between MWPT and ^{3}He export to Earth are also complex, with both competitive and collaborative considerations. We note that international regulations and safety concerns are potentially challenging to both ideas, but in neither case is there an obvious showstopper that we can see.

Both fusion and MWPT must be assessed against the 2040–2070 main time window of this book. We note that MWPT might be more viable in a shorter timeframe if limited to lunar power distribution locally.

The extended Lunar Energy Economy concept involves infrastructure sharing—a single large power source serving multiple clients, achieving economies of scale. This could be profitable if enough customers subscribe, turning energy into one of the first tradable commodities on the Moon (beyond raw materials). There is the prospect here of market structures that are not vertically integrated, facilitating the greatest number of market participants. A clear risk is that if the customer base is too small, the power provider cannot cover their fixed costs. This would probably lead

[4] By 'citadel' we mean a well-defined and secured compound able to sustain its own needs semi-indefinitely and without reliance upon vulnerable extended lunar networks.

to enhanced coordination (potentially even to a scenario where a user-consortium collectively funds a power station). This would provide stability, but it would weaken beneficial market forces.

The full vision of a networked and complex Lunar Energy Economy supporting diverse energy exports is a very bold idea. It takes us beyond the scope and time horizon of this book. We see it as something that might emerge after the propositions outlined in this book.

2.3.4 Proposition 4: Lunar Tourism—A Follow-on Development

We introduced the idea of space tourism earlier (In Chaps. 1 and 2 (Sect. 2.2)). We have considered whether such an economic driver be sufficient to motivate fusion on the Moon? Colvin et al. have looked at economic drivers associated with the Moon on a timescale out to 2040 [20]. Their work excluded military space considerations, and it considered both lunar and cis-lunar opportunities. Their conclusion (see Fig. 2.4) is that only space tourism has the potential to generate significant revenues on the 2040 timescale, but the issue soon becomes: *what does one mean by significant?* In Sect. 2.1, we argued that importantly much lunar development will occur before substantial commercial revenues start to appear. Such developments will be largely state-funded.

The amounts of money realistically available from such a proposition in isolation are not sufficient [20] to motivate ambitious lunar infrastructures like fusion on the Moon in the early phase of our period of interest. Looking beyond space tourism they caution that:

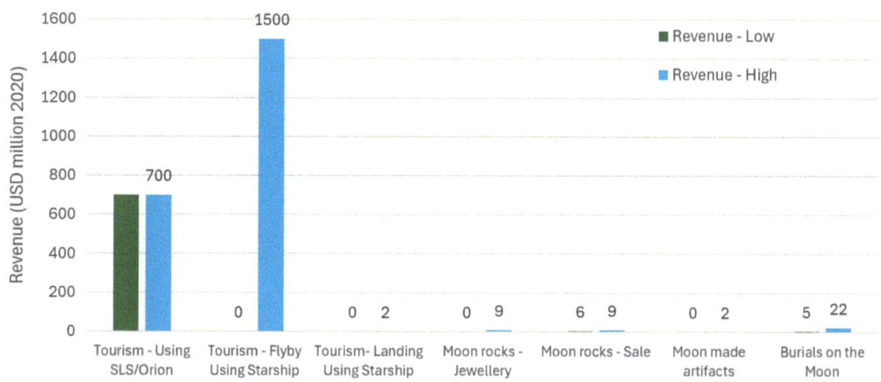

Fig. 2.4 Analysis by Colvin et al. for STPI of potential lunar commercial revenues in 2040. (Graph based on data from Colvin et al. ©2021 Institute for Defense Analyses, see [20])

> All other activities are uneconomic, for one or more of four reasons: the underlying technology is underdeveloped; there are no likely buyers in the 2040 timeframe; the cost of providing the services exceeds revenues obtained from selling it; or the product is cheaper to produce terrestrially than to produce on the Moon. In general, private demand is insufficient to sustain a pure-play lunar company. Such companies, if they are to become economically viable, need to find ways to exploit existing proven markets, either terrestrially or in near-Earth space.

They also note:

> Other than advertising, we found no good or service purchased by businesses that was economically viable in the timeframe of interest.

This relates to our discussion of branding in Sect. 2.3. We take note of these cautionary observations.

One might conclude that we are in disagreement with the conclusions of Colvin et al. but that is not the case. The key issue here is what scale of lunar development is one imagining? This book imagines a lunar future requiring, or enabled by, power at the MW level. At present such ideas represent an outlier scenario. Orthodox scenarios for the near future are at a much more modest scale of energy and infrastructure and it is in that context that Colvin et al.'s analysis makes sense. They argue that in a scenario of modest lunar development space tourism could be the most important and visible activity. We do not dispute the Colvin et al. conclusion within its frame of reference, but we suggest that in a future of rapid industrialisation, tourism will lag other developments. It is, however, because of the orthodox scenarios explored by Colvin et al. and others that we devote so much attention to space tourism in this book.

While the advertising proposition might be economically attractive to a business, we maintain the view that the revenues are simply insufficient to be relevant to fusion on the Moon. Figure 2.4 points to potential tourism revenues potentially as high as $1.5 billion per annum, but the costs will be similarly high, and we take the view that space tourism is unlikely (even by 2070) to play anything larger than a supporting role in scenarios for the lunar economy capable of utilising fusion on the Moon.

We further note that the consultancy PWC has suggested that, while space tourism may contribute to the lunar economy, sectors like transportation and resource extraction are more likely to be key drivers [21]. Our views align with such thinking, noting our focus on scenarios involving relatively high energy demand.

2.4 Commentary on the Four Propositions

In comparing the likelihood and viability of our four highlighted initiating propositions: We posit that helium-3 export could have the largest economic upside, driven by Earth's very large energy market. That proposition also has the clearest synergistic benefits with the choice of fusion on the Moon. We suggest that lunar settlement power is more of an enabling service (with indirect economic benefits). Such a service could be sustained using D–T fusion, but we suggest that would be on the basis of all fusion fuel being delivered from Earth. The power distribution services proposition could be an important competitor, or enabler, and currently its role is hard to assess. The tourism proposition is not irrelevant to fusion on the Moon, but we suggest it will only have a role in synergy with other propositions rather than leading an ambitious lunar development story.

It is likely that early lunar economic activity will rely on substantial government funding and not be immediately profitable. Over time, as more commercial operations develop (resource extraction, tourism, manufacturing), the private sector can find profitable niches in providing essential and desirable services.

The propositions presented here are not free-standing or exclusive. Each proposition reinforces the others. For example, a thriving helium-3 export business could justify improved lunar infrastructure (benefiting settlements), and a well-powered settlement makes it easier to scale mining, etc. If the scale of activity grows, and becomes safer, the tourism and true settlement might follow.

A holistic business strategy might involve a portfolio approach, where an entity (or coalition) invests in multiple aspects. For example, the company might mine helium-3 for Earth, use some of it to fuel a local fusion reactor to power a lunar base, and sell any surplus power or by-products to other lunar users or even back to Earth. In this way, revenue streams are diversified—some in Earth markets, and some in situ (improving overall project viability).

References

1. P.A. Berkman, President Eisenhower, the Antarctic treaty, and the origin of international spaces. Science Diplomacy. https://www.atsummit50.org/media/book-6.pdf
2. B. Tingley, We didn't know what it was at first. NASA aircraft uncovers site of secret cold war nuclear missile tunnels under Greenland ice sheet: LiveScience. https://www.livescience.com/space/we-didnt-know-what-it-was-at-first-nasa-aircraft-uncovers-site-of-secret-cold-war-nuclear-missile-tunnels-under-greenland-ice-sheet. Accessed Date 28 Mar 2025
3. Antartic Treaty Secreteriat, in *The Protocol on Environmental Protection to the Antarctic Treaty.* https://www.ats.aq/e/protocol.html
4. Arms Control Association, in *U.S.-Russian Nuclear Arms Control Agreements at a Glance.* https://www.armscontrol.org/factsheets/us-russian-nuclear-arms-control-agreements-glance
5. J. Rascon, in *NASA's Artemis Moon Landing Program Facing Cost Problems: Medi Talk; Updated* (2024, Jan 17). https://www.meritalk.com/articles/nasas-artemis-moon-landing-program-facing-cost-problems/
6. NASA, in *Commercial Lunar Payload Services.* https://www.nasa.gov/commercial-lunar-payload-services/. Accessed Date 29 Mar 2025
7. FT Film, in *Moon Rush: The Launch of a Lunar Economy.* https://www.youtube.com/watch?v=WCmO0fre9Fc
8. J. Foust, in *Musk Says SpaceX Focus is on the Moon Rather than Mars.* SpaceNews (2026, Feb 9). https://spacenews.com/musk-says-spacex-focus-is-on-the-moon-rather-than-mars/
9. S. Limbu, in *Amidst Growing Mountain Tourism in Nepal, Mount Everest Confronts the Perils of Pollution.* https://earth.org/amidst-growing-mountain-tourism-in-nepal-mount-everest-confronts-the-perils-of-pollution. Accessed Date 27 Mar 2025

10. S. Prasain, *Upper House Passes Tourism Bill with Tougher Everest Rules* (Kathmandu Times, 2026, Feb 2026). https://kathmandupost.com/money/2026/02/14/upper-house-passes-tourism-bill-with-tougher-everest-rules
11. GRU Space, in *Building America's First Moon Base & Hotel.* https://preview-client-project.b-cdn.net/GRU%20Space%20Whitepaper.pdf
12. C. Cookson, in *Will Space Tourism Really Lift Off?* Financial Times (2024, Oct 9). https://www.ft.com/content/4ec9715d-c151-4b91-ba1f-5c9bfb3363f7
13. J. Foust, in *The Search for a Commercial Lunar Economy: The Space Review.* https://www.thespacereview.com/article/4898/1#:~:text=That%20study%20projected%20the%20lunar,services%20accounted%20for%20the%20rest
14. NASA/JPL Visions of the Future (2020, Dec 24). Available at: https://www.jpl.nasa.gov/galleries/visions-of-the-future/
15. C.P. Spedding, W.J. Nuttall, S. Lim, Energy requirements of a thermally processed ISRU radiation shield for a lunar habitat. Adv. Space Res. **65**(11), 2467–2474. https://www.sciencedirect.com/science/article/pii/S0273117720301708 (2020). https://doi.org/10.1016/j.asr.2020.03.015
16. State of the Future, in *Nuclear Fusion: The State of Play* (2024). https://stateofthefuture.substack.com/p/nuclear-fusion-the-state-of-play
17. US Department of Energy, *Fusion Energy Strategy 2024* (2024). https://www.energy.gov/sites/default/files/2024-06/fusion-energy-strategy-2024.pdf
18. J.F. Santarius, in *Lunar ^{3}He, Fusion Propulsion, and Space Development.* https://ntrs.nasa.gov/api/citations/19930008235/downloads/19930008235.pdf
19. Orbital Today, in *Space Startup Interlune Wants To Dig Up The Moon For Helium-3: How Much Will It Cost?* https://orbitaltoday.com/2025/03/19/space-startup-interlune-wants-to-dig-up-the-moon-for-helium-3-how-much-will-it-cost/
20. T.J. Colvin, W.J. Crane, R. Lindbergh, B. Lal, *Demand Drivers of the Lunar and Cislunar Economy* (IDA Science and Technology Policy Institute, 2020). https://newspaceeconomy.ca/wp-content/uploads/2023/04/d-13219.pdf
21. PWC, in *Lunar Market Assessment: Market Trends and Challenges in the Development of a Lunar Economy* (2021, Sept). https://www.pwc.com.au/industry/space-industry/lunar-market-assessment-2021.pdf

Chapter 3
Technical Requirements of a Fusion Reactor on the Moon

Contents

3.1 Fusion Primer

At its simplest fusion is the application of Einstein's famous equation

$$E = Mc^2 \tag{3.1}$$

Two light nuclei are fused together to form a heavier nucleus. However, the mass of the resulting nucleus is slightly less than that of the two initial nuclei. This 'lost' mass is converted to energy. This book is not intended to give an extensive description of the science and engineering of nuclear fusion; details of these are given by Morse [1]. Similarly, readers interested in the commercialisation of nuclear fusion (on Earth) are referred to the work of Nuttall and colleagues [2]. There are a number of fusion reactions [1]; however, for the purposes of this book, only two will be considered in detail. These are based on deuterium (D), tritium (T) and ^{3}He. The two reactions are as follows:

(1) $\mathrm{D} + \mathrm{T} \rightarrow {}^4\mathrm{He} + n$ which releases 17.6 MeV per reaction,

(2) $\mathrm{D} + {}^3\mathrm{He} \rightarrow {}^4\mathrm{He} + \mathrm{H}$ with a release of 18.3 MeV.

These two examples are considered here as the first is the reaction most likely to be successfully employed on Earth (and subsequently the Moon), the second as this is believed to be the most likely candidate for large-scale employment of lunar ^{3}He.

W. J. Nuttall et al., *Fusion on the Moon*,
SpringerBriefs in Applied Sciences and Technology,
https://doi.org/10.1007/978-3-032-38217-7_3

Generally, the first reaction (D–T) has been favoured for fusion energy developments on Earth as it is the easier reaction to initiate. Broadly for a tokamak, D–T fusion requires temperatures around 50 million °C, while D–^{3}He fusion necessitates temperatures around 200 million °C. Temperature can be compensated for via another parameter (e.g. plasma pressure), but it is a clear truth that D–^{3}He fusion is significantly more difficult to achieve.

There is a ^{3}He-^{3}He fusion reaction, but it is more difficult that the two reactions shown above. We do not consider it further because we posit that for the Moon, it will always be easy to supply deuterium fuel from the Earth to the Moon, it will be the tritium that could be in short supply. As such, ^{3}He–^{3}He fusion on the Moon would be relatively wasteful of ^{3}He, and given the difficulty of the fusion process in that case, we do not consider ^{3}He-^{3}He fusion in any detail in this book.

Earlier, in Sect. 1.4, we pointed to the attributes of D-T fusion and the problems that arise when high energy neutrons are produced. The D-T reaction produces high energy neutrons, but the D–^{3}He reaction does not. It produces a charged particle. This can be steered with magnetic fields and even be used directly in power production.

Whatever the fuel—the nuclei involved in fusion reaction both carry a positive electrostatic charge and therefore will repel each other before they can fuse. To overcome this repulsion, it is necessary to give the nuclei sufficient energy, and to contain them for sufficient time, for a useful amount of energy to be released.

We noted earlier that the fusion reaction is not simply determined by temperature. At the risk of getting too deeply into the physics, for fusion to be sustained, the energy released by fusion must be equal to that lost from the system. This was used by Lawson to derive the 'Lawson Criterion' [3]. This states that for a sustainable reaction, in the case of deuterium and tritium at a temperature of 10 keV for a fusion reactor where Q, the ratio of the fusion power to the injected power is 2 and the efficiency of the conversion of the thermal power to electricity is 1/3.[1]

$$n\tau_{\varepsilon} \geq 10^{20}\ \mathrm{m}^{-3}\mathrm{s}, \tag{3.2}$$

where n is the number density of nuclei, and τ_{ε} the confinement time. The Lawson Criterion represents an important challenge for fusion technologies, such as tokamaks. It represents the threshold for net power production by the fusion device. A practical power generator would need to exceed the Lawson Criterion by a substantial margin, noting inevitable engineering system inefficiencies, in energy conversion etc. Nevertheless the Lawson Criterion represents an important technological criterion for assessing fusion technologies in development.

Generally, inertial approaches to fusion aim to increase the value of n, while magnetic confinement approaches aim to increase the confinement time τ_{ε}. In this document, only the magnetic confinement approach is considered, not because we do not see potential in other approaches, but rather to prevent this book becoming a description of fusion technologies already familiar to many of our readers. There are a range of approaches to magnetic confinement fusion. Examples include stellarators,

[1] A D-T fusion plasma at 10 keV has a temperature of approximately 116 million degrees Celsius.

tokamaks and mirror devices. In this book, we generally restrict consideration to a subset of tokamaks, namely the spherical tokamaks. In this approach, the plasma, rather than being in a 'doughnut' shape as in a conventional tokamak, is in a shape more akin to that of a cored apple. This provides a more efficient confinement of the plasma.

We tend to exclude stellarators for logistical and maintenance reasons. The coils in a stellarator are, to a first approximation, unique. In a tokamak, there is a much greater standardisation (and hence possibility for spares and backups). This taps into sectoral virtues in space policy terms, such as flexibility, resource efficiency and high levels of modularity.

3.2 Tokamak Fusion Reactors

In a tokamak, the fuel, either D–T or D–^{3}He, is introduced to the pressure vessel which has been evacuated to a vacuum. This mix of gases needs to be heated in order to become a plasma, i.e. the electrons and nuclei become disassociated. As originally conceived, a conventional tokamak uses a solenoid in the middle of the tokamak to induce a current in the gas. This leads to formation of the plasma and to a heating of the plasma. Radiofrequency electromagnetic waves are used to further heat the ionised gases using a gyrotron. Further heat can be introduced by the use of neutral beam injector in which a beam of uncharged particles (i.e. neutrons) is injected into the plasma where they transfer energy by colliding with the plasma. It must be acknowledged that neutral beam injectors are expensive pieces of equipment with significant power requirements.

Figure 3.1 shows the right-hand half of a tokamak in cross-section. In essence, the left-hand side would be a mirror image.

Because spherical tokamaks have limited volume in the centre, the use of a solenoid to generate a current, and thereby heat the plasma, is restricted. To overcome this, some spherical tokamak designs incorporate 'merging compression' [4]. In this approach, two plasma rings are formed using two poloidal field coils. These are then merged via magnetic reconnection into one plasma ring that then is radially compressed, thus raising the temperature.

When the plasma becomes sufficiently hot, the fusion reactions start to occur, and the plasma becomes self-heating. If the plasma were to make contact with the wall of the vacuum chamber, then the plasma would lose energy, and the wall could potentially be damaged. Thus, the plasma needs to be contained and kept away from the walls. As the plasma is composed of charged particles, it can be confined by magnetic fields. One set of coils generate an intense toroidal field while a central solenoid creates a poloidal field, confining the plasma particles. A third set of magnetic coils creates an outer poloidal field to shape and position the plasma within the torus. To produce the high magnetic fields required, the magnets will be constructed from

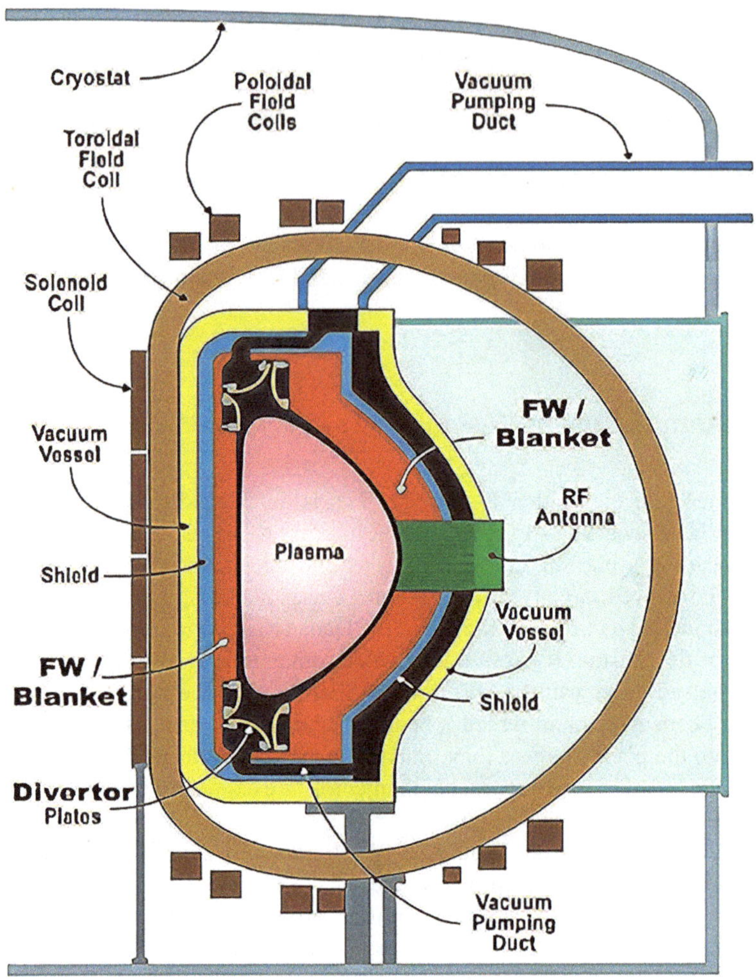

Fig. 3.1 Main components of a Tokamak fusion reactor. *Image Source* Şahin, S. (2019). Selection criteria for fusion reactor structures. *Journal of Thermal Engineering*. https://doi.org/10.18186/thermal.531703 Creative Commons Licence CC-BY-4.0

superconducting materials.[2] The orientation of the main toroidal and poloidal fields is illustrated in Fig. 3.2.

Whatever method is used to heat the plasma prior to it becoming self-sustaining, a large amount of energy will be required. For example, ITER will require electrical power in the range 110–620 MW for a period of up to 30 s during start-up [5]. It will be important for all fusion applications that the power requirement is minimised.

[2] For information on ideas relating to the use of cryogenic liquid hydrogen at approximately 20 K as a coolant for HTS magnets, we recommend that readers perform internet searches on the term 'hydrogen cryomagnetics'.

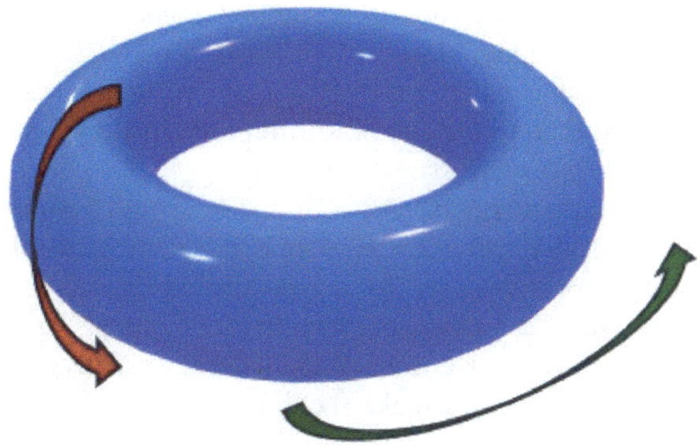

Fig. 3.2 The toroidal (green) and poloidal (red) magnetic fields interact to produce a closed-loop helical magnetic field which confines the plasma. Authors' own work

That will be especially important for space applications where energy supply will, in the absence of fusion, be very limited. A key concern will not just be to reduce peak power requirements at start-up, but also to reduce the amount of time over which start-up power is required. Hence, in effect, the key challenge is to reduce the electrical energy required rather than the peak power. This requirement means that it will be necessary for there to be a suitable source of energy already present on the Moon and for the tokamak facility to include suitable storage and discharge capability. As described previously (Sect. 1.5), part of the lunar facility might be an array of solar photovoltaic cells coupled to an accumulation system sufficient for short duration, but high power, discharge.

Noting the energy needs of ITER above, we must point out that ITER is a part of a traditional and large-scale vision for tokamak energy. In this book, we focus on the innovation of spherical tokamaks (STs) using high-temperature superconducting (HTS) magnetic coils. The ST approach brings with it various advantages that might be hugely beneficial to fusion on the Moon. Key advantages include smaller physical size and the prospect of lower start-up energy requirements.

3.3 Fuel Supply and Helium 'Ash' Removal

As the fuel is consumed in the fusion reaction, these will need to be replenished. This is actually a significant safety attribute of fusion over fission. There is simply very little stored energy in a fusion reactor at any given point. A fission reactor contains within it enough fuel (energy) for many months of operation.

On Earth, the main candidate fusion reaction is D–T. Deuterium is present as a small proportion of all hydrogen. Given the relatively small quantities involved, it would be feasible to upmass deuterium to the Moon. We suggest further research to double-check this assertion.

On Earth, tritium fusion fuel, apart from that used for initial operations, will be produced in a breeder blanket in which tritium is formed as a result of neutron capture reactions. The tritium formed, together with ^{4}He, needs to be removed from the blanket material and extracted in a recovery plant. The breeder elements are massive items which together with the tritium extraction plant increase the mass to be transferred to the Moon if tritium production were to be established in situ from lunar D–T systems (see: Sect. 1.5).

Lunar deployment of D–^{3}He fusion would have no need for problematic tritium supply from Earth or for even more problematic local production. The use of D-^{3}He fusion avoids the need for a breeder blanket and it would also reduce the requirements on in vessel remote handling equipment. Note also that we regard the local lunar supply of ^{3}He to be abundant and in all circumstances D supply from Earth should be easy. We make the claim because the mass of fusion fuels is low. For example, calculations carried out by the UKAEA [6] have shown that a D–T fusion reactor would require 0.056 kg of tritium to produce 1 MW-year of energy. Similarly, 0.037 kg of deuterium would be required for the same output. Even with the inclusion of additional mass connected with the packaging and chemical forms in which the deuterium and tritium are transported (noting safety concerns in the case of radioactive tritium), the fusion fuel upmass is likely to form only a small component of the total upmass required to supply a lunar settlement or industrial operation.

In the D–T fusion process, ^{4}He is produced. This ^{4}He, if it remains in the chamber, will dilute the plasma and impede the reaction. To prevent this, the helium 'ash' and other impurities are removed by configuring the poloidal magnetic field so that the edge of the plasma impinges on a divertor, usually at the bottom of the tokamak, which collects and removes these impurities. If, however, the fusion vessel were to be vented directly from the chamber (with a system to retain the deuterium and tritium for reuse), then the challenge of removing the plasma-cooling fusion 'ash' might be reduced. In extremis—the removal of this requirement would allow the use of simpler divertor designs which would only need to be employed for heat removal. We note that, on the Moon, venting is easy, although it is important to stress that the post-burn contents of a tokamak vacuum vessel would contain very valuable gases. The low ambient temperatures at the lunar poles might permit relatively easy cryo-pumping and gas regeneration and recycle. In such a scenario, only true wastes would be vented to the vacuum of space. The duty cycle of a fusion tokamak on the Moon is an issue meriting further investigation and research.

Compared to fusion on the Earth, in some important ways, our proposal is for a greatly simplified fusion reactor design—we shall return to this point in Sect. 5.2.

References

1. E. Morse, *Nuclear Fusion* (Springer, 2018). ISBN: 978-3-319-98172-2
2. W.J. Nuttall, S. Konishi, S. Takeda, D. Webbe-Wood, *Commercializing Fusion Energy: How Small Businesses are Transforming Big Science*, 2nd edn. (Porto Press, 2025). ISBN 9781849956048

3. J.D. Lawson, Some criteria for a useful thermonuclear reactor, in *Proceedings of the Physical Society* (1957) https://www.euro-fusion.org/fileadmin/user_upload/Archive/wp-content/uploads/2012/10/dec05-aere-gpr1807.pdf
4. P.F. Buxton, M. Gryaznevich, Tokamak Energy Ltd Team, Merging compression start-up predictions for ST40. Fusion Eng. Des. **123** (2017). https://doi.org/10.1016/j.fusengdes.2017.04.067
5. ITER, Power Supply. https://www.iter.org/machine/supporting-systems/power-supply. Accessed Date 12 Mar 2025
6. S. Zheng, D.B. King, L. Garzotti, E. Surrey, T.N. Todd, *Fusion Reactor Start-up without an External Tritium Source.* Contract No.: CCFE-PR(15)68 (2025). https://scientific-publications.ukaea.uk/wp-content/uploads/Preprints/CCFE-PR1568.pdf

Chapter 4
The Lunar Environment

Contents

4.1 Siting Considerations

The lunar terrain, characterised by its irregular topography, abrasive regolith, and harsh and long diurnal cycles, presents significant logistical barriers to the deployment of sensitive fusion technology. This is in addition to the challenges of landing spacecraft in the mountainous landscapes expected for PSR.

Suitable locations for a fusion plant and for human habitation are likely to be found in lunar caves similar to lava tubes found on Earth. Such a tube, on Lanzarote, is shown in Fig. 4.1.

The transfer of tokamak components from lunar surface access points to designated PSR locations will require specialised—or perhaps, conversely, generalised—robotic systems. These systems must possess advanced navigational and manipulative capabilities to traverse the variable terrain while protecting the delicate integrity of the equipment. Temperature differentials between the equipment and the lunar environment demand sophisticated thermal control mechanisms during transport.

We envisage that there will be lunar workshops capable of major systems maintenance at least at the level of tokamak toroidal field coils and perhaps at the level of whole vacuum vessels. As regards the transport of engineered items across the lunar surface, care will be needed to minimise regolith disturbance. The risk of exposure

W. J. Nuttall et al., *Fusion on the Moon*,
SpringerBriefs in Applied Sciences and Technology,
https://doi.org/10.1007/978-3-032-38217-7_4

Fig. 4.1 Lava tube at Geoparque Lanzarote. *Source* https://www.esa.int/ESA_Multimedia/Images/2016/11/Lava_tube. Credit ESA–L. Ricci

to the near-ubiquitous and electrostatically charged regolith dust requires the implementation of robust mitigation protocols to safeguard sensitive components from contamination and abrasion. The need to clean high-precision items in a mine works location is something best avoided.

4.1.1 Repurposing and Modularity

One important aspect of space innovation has been a pragmatic willingness to re-use and improvise. Thus far, the fusion industry on Earth has arguably not thought in such terms. We expect system reliability to be a growing concern for terrestrial fusion developers, especially given their persistent focus, in some quarters at least, on baseload electricity generation. One aspect of reliability is the ability to perform maintenance quickly. The isolation of space endeavours and the wider challenges of upmass will only amplify such thinking around reliability and resilience. The space community is ahead of the fusion community in this regard. For space engineers, the virtues of self-sufficiency and resilience are manifest.

A key part of such notions of efficiency is the benefit of reuse and repurposing of space quality items designed and approved for other purposes. Perhaps the clearest example of such a repurposing mindset, as relates to normal operations at least, is the Skylab space station of the 1970s. The SkyLab structure was formed of a repurposed a

Saturn V second stage fuel tank from the Apollo programme. In emergency situations, the same improvisational space mindset can also be seen in space engineering. The famous Apollo XIII accident and the guided safe return of the astronauts relied upon an unanticipated repurposing of the lunar lander module as a lifeboat to get the astronauts back to Earth orbit. With such lessons in mind, we suggest that fusion, with its requirement for strong lightweight vacuum vessels, could make use of modular repurposing.

Such ideas could not be expected to apply in the early stages of exploration and settlement but could form part of later thinking once the settlement is firmly established. Rather than saying that spacecraft structures will be specially reused for fusion, we actually posit that they will be used for numerous pressurised purposes, for industry and the settlement. Such modular items could be expected to arrive frequently from Earth. A proportion would of course make the return trip (e.g. for crew rotation) but not all. Many more craft will be travelling from Earth than will return. It is not just human beings that need to be transported with life support (plants, micro-organisms for research and even luxury fresh food items) might all travel in pressurised unmanned cargo modules.

4.1.2 Regolith Accumulation

As noted in the discussion of equipment transfer on the lunar surface, low-gravity, electrostatic forces, and the high vacuum environment will conspire to accumulate abrasive regolith dust in ways that will be deleterious to the operation of the tokamak and its ancillary equipment. Lunar regolith, characterised by its abrasive properties, presents a persistent contamination risk to fusion equipment. Dust generation occurs during various operational phases, including venting, transport, setup and maintenance. Rocket exhaust debris also contributes significantly to dust accumulation. Indeed, it can also measurably alter the lunar atmospheric pressure (although it is always UHV).

The electrostatic charging of dust particles, a unique characteristic of the lunar environment, can lead to adherence to surfaces, greatly exacerbating contamination issues. The implementation of effective mitigation strategies, such as electrostatic dust shields and mechanical cleaning systems, is essential to minimise contamination. The performance of these strategies must be evaluated under simulated and actual lunar conditions to ensure their efficacy in protecting critical equipment.

Tokamak fusion uses large magnetic fields—we need to be careful as to how this will affect the behaviour of dispersed regolith dust in the work environment.

4.2 Ambient Temperatures on the Moon

In this book, we consider the properties of the permanently dark regions at the base of craters at the lunar poles. There remains some uncertainty about the surface and bedrock temperatures in these locations. Nasa's Diviner Lunar Radiometer Experiment [1] has been mapping the lunar surface since June 2009. Williams et al. report on their findings for two craters near the lunar south pole:

> … temperatures on the floor of Faustini vary ~30–40 K in winter, while the range is ~45–70 K in summer. The floor of Haworth, in addition to being one of the coldest PSRs, experiences smaller variations in temperatures: ~20–30 K in winter increasing to ~30–45 K in summer [2].

As far as we are aware, bedrock temperatures in these polar PSR regions remain unknown, but for the rest of this book, we shall assume, perhaps optimistically, that cryogenic temperatures consistent with the use of high-temperature superconductivity will be available in the local geology. We further posit that it will be possible to access and manage this cold source in ways that do not cause it to warm unacceptably. For the rest of this book, we shall refer to an ambient 50 K bedrock cold source, but our statements must be treated with some caution given the uncertainties involved.

Williams et al. also interestingly observe that:

> Areas cold enough to trap water ice during the winter season are significantly larger than the cold, permanently shadowed regions [2].

We conclude that for the purposes of fusion on the Moon, as described in this book the available locations could be few and very precious if the ideas that follow are to be viable in full. To summarise: At its most extreme, our thinking imagines a fusion tokamak operating within the lunar PSRs, where in certain special locations, ambient temperatures may be available that are sufficiently low to operate HTS coils without active refrigeration. Of course, the tokamak itself will be a major heat source (MW)—that heat must not be allowed to contaminate the cold bedrock.

Operating a tokamak in a very low-temperature environment requires investigation of the fundamental behaviour of fusion engineering materials under extreme cryogenic conditions. Thermal contraction, a non-linear phenomenon, necessitates meticulous material selection and design. The varying coefficients of thermal expansion (CTEs) of constituent materials demand a system-engineering approach to prevent stress-induced structural failures. This entails advanced modelling and analytical techniques to forecast and accommodate dimensional variations across the tokamak's intricate architecture, encompassing superconducting magnets, vacuum systems and structural supports. Determining the impact of design and manufacture of components in an environment 250 K warmer than its operating environment needs to be understood. Lessons from the James Webb Space Telescope (JWST) and other satellites and rovers operating in space must be absorbed before the design phase.

The logistical demands of transporting the tokamak (or its component parts for lunar assembly) to the lunar surface, and subsequently to its operational PSR location, introduce further complexities in thermal management relating, for example, to structural engineering. The thermal cycling associated with testing, transportation and operational phases require comprehensive material characterisation and system-level validation. The dynamic interplay between cryogenic conditions and material properties forms a core consideration in the feasibility assessment of this endeavour.

The thermal conductivity of lunar regolith, a poorly characterised medium, and its interaction with cooling systems requires detailed analysis. For this reason, we envisage the tokamak firmly fixed to, and embedded in, high thermal conductivity PSV lunar bedrock at cryogenic temperatures (circa 50 K). The absence of convective heat transfer in the lunar vacuum means total reliance on radiative and conductive heat transfer regimes. The design of the energy conversion (heat to electricity) system must consider the specific thermal properties of materials and the operational parameters of the tokamak. The dynamic interactions of these thermal factors dictate the overall engineering design and operational efficiency.

It is important to note that the question of how to survive the temperatures in the lunar environment extends beyond the tokamak itself to the power transmission system used to deliver the power to demand nodes. Cabling systems must be constructed from materials that retain conductivity and flexibility at 50 K. We note the opportunities afforded by today's high-temperature superconducting (HTS) rare-earth-based REBCO tape-based conductors. We also note that truly flexible wires can be made from the superconductor MgB_2, although in that case the lunar context may be too warm for effective use without extra cooling. If ambient temperature superconductivity cannot be deployed, and our hope would be that it can be used for at least part of the cable run, then the length and thickness of cables influence power losses and signal integrity, requiring careful optimisation. We note that for pulsed current aspects for specific tasks, copper is an excellent low-loss conductor at cryogenic temperatures.

Insulation and shielding materials must be compatible with the lunar environment. That environment will give rise to operational parameters relating to long-term reliability in the face of the high charged particle flux on the lunar surface. We see such benefits of deploying cabling in conduits drilled through the cold bedrock that we tend to recommend it as an option despite its energy needs during construction. Eventually, if we achieve fusion on the Moon, we will have energy in abundance. The first phase of development will, however, clearly involve some difficult engineering trade-offs and some careful energy management planning. The professional skills and techniques of civil engineering and construction will be essential for the success of any fusion on the Moon scenario.

In our preferred HTS plan—linking to an ambient cryogenic environment—electrical connectors must maintain their integrity, guaranteeing reliable and resilient performance. As with the Tokamak, the design of the electrical system must account for thermal contraction and mechanical stress, ensuring structural and functional integrity. Material selection and component testing are essential to validate the performance and reliability of the electrical systems in the extreme lunar environment.

References

1. J.P. Williams, D.A. Paige, B.T. Greenhagen, E. Sefton-Nash, The global surface temperatures of the moon as measured by the diviner lunar radiometer experiment. Icarus **283**, 300–325 (2017). https://www.sciencedirect.com/science/article/pii/S0019103516304869. https://doi.org/10.1016/j.icarus.2016.08.012
2. J.P. Williams, B.T. Greenhagen, D.A. Paige, N. Schorghofer, E. Sefton-Nash, P.O. Hayne, et al., Seasonal polar temperatures on the moon. J. Geophys. Res. Planets **124**(10), 2505–2521 (2019). https://doi.org/10.1029/2019JE006028

Chapter 5
Lunar Industrial Proposition—In Detail

Contents

In this Chapter we develop and explore one of proposed scenarios for the development of Fusion on the Moon. These propositions were introduced in Chap. 2 in Sect. 2.3. In this chapter we shall focus on the proposition introduced in Sect. 2.3.1 where it was described as “Helium-3 Mining and Export to Earth”. In this Chapter we shall widen our thinking slightly and take a more technical approach. We here consider the Lunar Industrial Proposition.

We return to the idea introduced in Chap. 2 that the reactor is such a proposition should be co-located with three key resource opportunities: water-ice, helium-3 and metallic deposits. The metallic deposits of interest will be from asteroids and meteoroids that impacted the lunar surface perhaps millions of years ago. On Earth, biological activity and water chemistry would be expected to erode such a metallic object over millennia.[1] Over longer time periods, plate tectonics and the deposition of soils and other geological material would bury what was left of the original object. Basically on Earth, ancient high metal content deep space objects are not available to us, but they are on the Moon. There is no biological activity, there is little or no water chemistry (noting that even if there is water it is at cryogenic temperatures) and there is no geophysical activity. So where does one find high purity metals on the Moon?—in the middle of ancient impact craters. All one needs do is dig down a

[1] In this connection, it is noteworthy that the discovery in 1925 of a 3400-year-old iron dagger in Tutankhamun’s tomb was a source of mystery as the date came far earlier than mankind’s ability to produce metallic iron. The truth was later confirmed via X-ray analysis that the iron in the dagger had been discovered in ancient times as a metal. It had come from space as a meteorite.

W. J. Nuttall et al., *Fusion on the Moon*,
SpringerBriefs in Applied Sciences and Technology,
https://doi.org/10.1007/978-3-032-38217-7_5

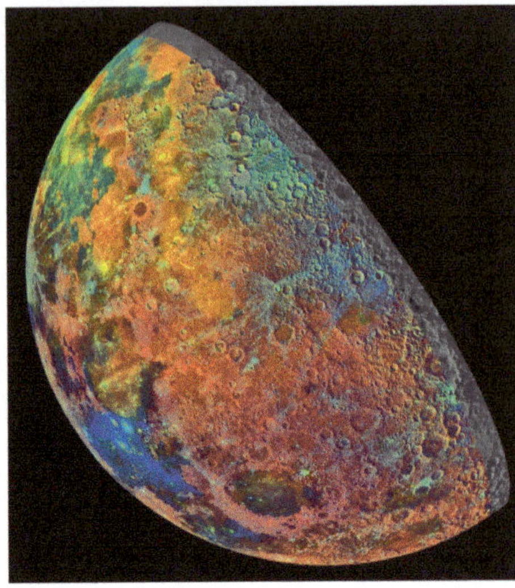

Fig. 5.1 Artificial colour representation of the diversity of minerals on the lunar surface. *Source* NASA/JPL

relatively small distance. Looking at water, helium-3 and metals we are drawn to one location above all others—polar craters. This is fortunate as they are already widely regarded as interesting to the lunar surface actors who might embark on a mission to deploy a fusion reactor on the Moon. Our ideas fit with much larger plans already under development. We anticipate user-pull for our ideas rather than this being a story of technology push.

In Fig. 5.1, we reproduce an artificially coloured photographic montage of the northern areas of the Moon showing diverse surface minerals. The underlying data was obtained through the use of spectroscopic filters by the European Galileo probe on 7 December 1992. It points to the mineralogical diversity of the lunar surface.

In terms of specifics, shielding of any human or robotic crew from ionising radiation produced by the fusion reactor is the critical factor (especially if it is a D–T reactor), and therefore, it is posited here that the reactor is sited at the end of a tunnel bored out from the side of a crater floor, with the human habitation and sites of activity similarly dug out, at an appropriate distance, given the material's radiation attenuating properties (see Fig. 5.2).

5.1 The Industrial Facility

The reactor element of this proposition cannot easily be separated from its function or its purpose on the Moon, and as such, we provide an indicative look at the activities undertaken at, and the particulars of, the wider lunar settlement, of which the reactor is part of this proposition.

Fig. 5.2 Schematic of a fusion powered lunar mining activity. Note physical separation of Tokamak and human habitation. *Image source* T.B. Scott (with permission, all rights reserved)

5.1.1 Habitat and Life Support Functions

The manufacture and operation of a lunar habitat fit for a sustained human presence are evidently sequential. The habitat must be built before it can be occupied. However, the building of the habitat requires substantially more power and energy than the operation of it. Massive amounts of shielding are required to ensure that even the mildest of solar storms does not immediately kill all humans on the surface of the Moon, and either digging, or sintering large-scale habitats from the local terrain and regolith are extraordinarily energy intensive—on the order of tens to hundreds of megawatts [2]. Base operations, meanwhile, could reasonably be expected to require only a few hundred kilowatts or less, dependent on crew numbers, rotations, and other marginal loads. As such, if the reactor is only required for a build phase, then it becomes an extremely expensive way of providing operational power. Therefore, our position is that there must be ongoing and energy-intensive activities nearby to justify the design, build and launch of a fusion reactor for lunar surface power. We further suggest that the first enabling technologies for lunar construction in PSR could be fission-based. That technology can deliver substantial power reliably through the lunar night, but it comes with political concerns that will not get easier. The alternative is to establish solar PV capability during the polar summer—establish fusion capability on that basis and then, making use of the energy accumulation capability inherited from the solar PV rig the fusion tokamak might reliably pulse its way through the polar lunar winter and, beyond.

5.1.2 Helium-3 Mining

Helium-3 mining, both for self-sufficiency and economic export, is an energy-intensive industrial lunar activity, and given its critical juncture with the fusion reactor itself, we suggest it could be the first such industrial activity, and indeed the first lunar activity of any type at that scale (personnel, upmass and invested capital). In the early phase, it will be important to prove out the technologies and practices required to mine effectively and to export economically relevant materials from the Moon. Those subsequent export resources could be:

- *Water-ice, processed for astronaut consumption* (e.g. *for onward missions to Mars); for electrolysis for rocket fuel; or some other as yet unknown industrial process.*
- *Meteoritic metals, deposited in high concentrations in the centre of craters formed by metal-rich impactors.*

With that said, it could be that the only material ever to achieve lunar export value could be ^{3}He. The shallow depths at which helium-3 is found on the Moon means that scraping regolith powder from the surface should be sufficient. We, and others, anticipate meaningful amounts of helium-3 to be found deposited in just one tonne of the bulk material. Once a rover had collected enough of the material, it would be necessary to process it for volatiles, and pipe it through to the reactor location (buried into the side of the crater (see Fig. 5.2)). Figure 5.3 illustrates a harvester concept being developed by Interlune [3].

Fig. 5.3 Interlune harvester concept. *Image source* Interlune, see: https://www.interlune.space/press-release/space-resources-company-interlune-unveils-full-scale-prototype-of-excavator-for-harvesting-helium-3-from-the-moonH

If initially nuclear energy for industrial use is limited, then Fresnel lenses or other solar collection approaches could be used (especially in summer) to achieve temperatures (many hundreds of degrees C) sufficient to extract 45% of the ^3He. Oxygen production, via the solid-state processing of regolith, can also be achieved at this stage, as well as any more complex mineralogical extractions or beneficiations.

Text Box 4—The year 2051—we imagine ...

Embedded in the extremely cold bedrock (approximately 50 K) in a permanently dark lunar crater at the lunar south pole one finds a fusion tokamak reactor. The mining operations that it supports have been running for 6 years and the fusion reactor runs on D–T fuel supplied from Earth. The mining operation has been sending ^3He back to Earth for two years now. The main customers have been those developing ever larger ultra-cold quantum computers. Fusion uses have been under development for many years, since the Moon started to open up in the late 2020s. Next month it is expected that the first terrestrial tokamak will use lunar ^3He in test operations. None of the revenues that have so far materialised (and this business is definitely already in revenue) have yet had anything to do with direct energy uses of ^3He. In early 2051, industrial interest is mostly focussed on a different set of tests, also starting next month, ^3He to T conversion on Earth. Investors are, of course, eagerly awaiting the results of both tests, which are being led by one company—the clear market leader in the field of lunar fusion fuel cycles. The business press says that the company's share price has one success, and one failure priced in. Two failures would be very bad news.

Looking ahead to D–^3He fusion, the company already has engineering designs for a D–^3He lunar tokamak that might be deployed onsite at the lunar south pole in 2056. Geopolitical rivalries are intense on the Moon, but the Guardians of US Space Command recently established a permanently staffed facility 3 km away from the company's crater. The United States is supporting the company's international assertions that it now, in effect owns the crater and all its mineral wealth, although the government still talks in terms of 'safety zones', rather than anything that directly suggests 'ownership'.

5.2 Lunar Tokamak Concept—A Schematic

In Fig. 5.4 we provide a schematic of a tokamak facility on the Moon. With reference to that figure we note that:

- The tokamak facility is embedded in the dark bedrock of an excavated cavern at ambient cryogenic temperatures (see Fig. 5.2).

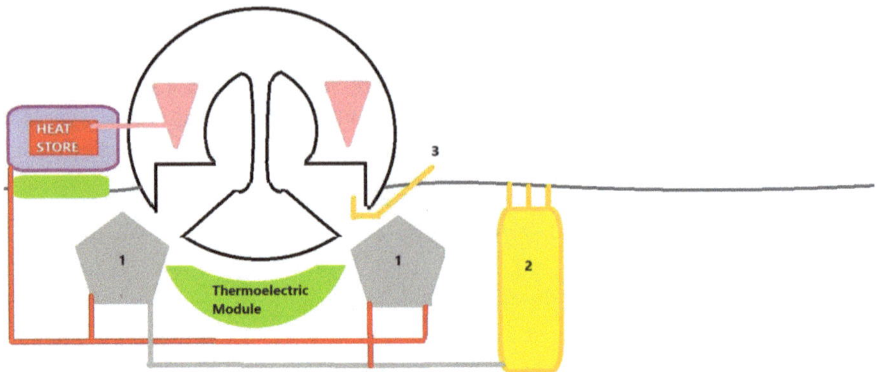

Fig. 5.4 Schematic of the Lunar Tokamak constructed on lunar bedrock. *Source* Authors' own work

- The facility makes extensive use of approx. 50 K ambient HTS cabling. There is a ^{4}He cryogen loop deep into the cold near-surface bedrock for main tokamak coil cooling.
- The facility has no breeder blanket—all fuel is supplied to site.
- D–T neutrons are thermalised in an improvised (molten salt/regolith) blanket—shown in pink.
- Heat is stored in a heavily insulated heat store at 700 K.
- Heat is transferred using molten salts.
- 700 K system heat and the cold bedrock are handled at the base of the device.
- The system has a 4 square metre footprint with an approximate 650 K temperature difference.
- The temperature difference drives an array of thermoelectric generators (green).
- As noted earlier, it is important that the tokamak in operation does not contaminate the cold bedrock with waste heat. Some level of refrigeration-free rock cooling may be needed to maintain thermal gradients below the tokamak.
- The coldest, and most precious, cryogenic supply will be provided to the magnet coils.
- Item 1 is a divertor-adsorber—essentially a cryopump—it is cooled with the He return gas from the magnet coils.
- Item 2 is a gas reprocessing system—for tritium, helium and deuterium recovery at the end of each fusion pulse.
- Item 3 is a direct vent port—to ambient Lunar UHV for system final clean prior to recharge for next pulse.

The extreme cold of permanently shadowed lunar regions, where ambient temperatures can reach the low tens of Kelvin, allows for the efficient use of high-temperature superconducting (HTS) materials critical for magnetic plasma confinement. The enhanced phase diagram margins of HTS materials at these temperatures translate to higher critical currents and critical fields, enabling the generation of stronger magnetic fields without significant risk of a magnet 'quench'.[2]

The very cold bedrock of the Moon in PSR locations also permits an attractive power generation opportunity free from rotating machinery and high-pressure fluids. Thermoelectric power generation (such as the thermionic systems used in today's radioisotope thermoelectric generators) has historically been limited by two considerations. The temperature gradient accessible using radiative coolers (as on NASA Mars rovers) and the surface area across which there is a temperature difference. Our proposition allows for a very large surface area of contact between the tokamak and the bedrock.

It is important to recognise that our hot blanket is not a fuel breeder. Engineering optimisation is all about heat and power, not the fuel cycle.

Thermoelectric generators are high reliability solid-state devices with no moving parts. Given the challenges associated with maintaining turbomachinery in the lunar environment, conventional steam cycles are liable to require more maintenance than solid-state thermionic electricity generation, akin to the technology used in RTGs, coupled with the substantial thermal gradient between reactor and heat sink, thermionic power generation may be the most efficient long-term solution, if they can be scaled up sufficiently. There may also be an opportunity to explore the poten tial of gamma voltaics to convert the gamma radiation produced by the D–T fusion reaction directly into electrical power.

Even before the later move to D–^{3}He fusion, ours is a power system without:

- *Cryogenic refrigeration.*
- *Fuel breeding.*
- *High-capacity vacuum pumps.*
- *Rotating turbomachinery.*

The tokamak is much simpler than that required for fusion on the Earth—it is that simple observation of the potential for fusion on the Moon that has, in part, motivated the authors to prepare this book.

Managing the thermal output of the fusion reactor and providing heating for the lunar habitat are interconnected considerations. The ultra-high vacuum of the lunar surface (approximately 10^{-15} bar) eliminates convective heat transfer, making radiant heat transfer the primary heat transfer regime, with concern given also to the conductive heat sink of the bedrock, along with thermal management to ensure life support systems for the crewed areas.

---

[2] An HTS magnet is said to quench when it loses superconductivity. When part of the magnet becomes a normal conductor, there is an electrical resistance and the local current causes heating which causes adjacent HTS material to stop being a superconductor. In a very short period, the whole magnet becomes a normal conductor, and the heat generated causes the cryogenic coolant to vaporise. HTS magnets are designed with such risks in mind, but quenches are to be avoided.

The dedicated heat storage system will be crucial for buffering fluctuations in reactor output and ensuring a consistent heat supply, especially during maintenance periods. The design of this storage system, along with the insulation strategies employed, will be paramount for the long-term sustainability of the lunar power plant.

The operational profile of the reactor is envisioned as a pulsed duty cycle, with a plasma burn of approximately 20 min followed by controlled gas extraction for fuel reprocessing and subsequent venting of the main chamber to the lunar vacuum. This mode of operation allows for efficient fuel management and waste disposal. The lunar vacuum offers a significant safety advantage, as emergency venting of the reactor chamber poses no threat to personnel. Heat generated by the pulse would be stored as shown in Fig. 5.4.

The structural demands on a lunar tokamak differ from those on Earth. While the reactor will experience significant thermal cycling due to its pulsed operation, the reduced lunar gravity mitigates the mechanical stresses on the structure, and a heavy breeder blanket is avoided, substantially reducing the reactor's overall mass (and thus any thermally induced creep stresses). This all simplifies the structural requirements, but we acknowledge the specific special structural challenges arising from the lunar environment as discussed earlier.

Several key research questions need to be addressed to realise a lunar tokamak fusion reactor:

- Determining the minimum infrastructure required for heating both D–T and ^{3}He-fuels to fusion conditions in a lunar environment.
- Those lessons will be essential for optimising reactor design and minimising mass.
- System reliability will be a key concern.
- The optimal duty cycle, balancing plasma burn duration with reset time, needs further investigation to maximise power output while minimising thermal stress on the reactor structure and ensuring system end-use effectiveness.
- While reduced gravity alleviates some concerns regarding creep fatigue, a thorough analysis of the structural response to repeated thermal and magnetic cycling is necessary.
- Finally, a comparative assessment of tokamak versus stellarator designs for a lunar fusion reactor is warranted. While stellarators offer the potential for continuous operation, avoiding thermal cycling issues, they typically lack the redundancy inherent in tokamak designs, which could be a critical factor for reliability in a remote lunar setting.

Finally in this book, we focus on the attributes of magnetic confinement fusion, but we do not intend to exclude pinch or mirror concepts or indeed any other ideas that might suit the lunar environment.

References

1. NASA. https://science.nasa.gov/photojournal/moon-false-color-mosaic-2/
2. C.P. Spedding, W.J. Nuttall, S. Lim, Energy requirements of a thermally processed ISRU radiation shield for a lunar habitat. Adv. Space Res. **65**(11), 2467–2474. https://www.sciencedirect.com/science/article/pii/S0273117720301708 (2020). https://doi.org/10.1016/j.asr.2020.03.015
3. https://www.interlune.space/about. Accessed Date 13 May 2026

Chapter 6
Uncertainties Surfaced by the Lunar Industrial Proposition

Contents

6.1 Information and Governance Uncertainties

Foundational to any lunar mission is the current lack of high-fidelity data regarding the local environment and its regulation. There are significant gaps in our knowledge of the local abundance of deuterium, He, lithium and water–ice, as well as the mechanical and thermal properties of the regolith. This is compounded by the variance between satellite observations and actual ground truth. Legally, the framework for these activities remains unsettled, specifically regarding the status of the Moon Treaty at the time of launch and the precise requirements for decommissioning lunar facilities and managing radiological waste. While territorial claims to the Moon are blocked by the Outer Space Treaty, we are not aware of any credible moves underway that could block lunar resource extraction. This is, however, an issue meriting ongoing attention.

6.2 Economic and Market Uncertainties

The commercial viability of lunar mining depends on speculative market dynamics and long-term infrastructure choices. The potential demand for lunar-sourced He and fusion energy is tied to anticipated terrestrial prices in the 2040–2070 period,

W. J. Nuttall et al., *Fusion on the Moon*,
SpringerBriefs in Applied Sciences and Technology,
https://doi.org/10.1007/978-3-032-38217-7_6

which remain difficult to forecast. We must determine whether the export of these resources is economically justified compared to orbital alternatives and whether a lunar spaceport is a credible centrepiece for a self-sustaining ecosystem. Additionally, the environmental impacts of lunar activities pose a risk to commercial sustainability, and we here adopt the assumption that space tourism will not be a significant driver for fusion. That assumption merits further validation.

6.3 Operational and Design Uncertainties

Sustaining a fusion-based presence on the Moon introduces complex operational and engineering requirements. On the design front, uncertainty remains regarding the necessity of breeder blankets, methods for storing reactor pulse energy and the selection of appropriate launch systems. Operationally, the mission must account for the required level of maintenance automation, common failure modes and the logistics of buffer stores for consumables. Furthermore, the industry must validate the actual need for megawatt-scale energy in permanently shadowed regions (PSRs) and ensure all reactor siting meets stringent STEEPLE (Social, Technological, Economic, Environmental, Political, Legal and Ethical) requirements.

6.4 Technical and Environmental Uncertainties

The densest category of uncertainty involves the physical and logistical reality of operating complex machinery in deep space. Technical challenges range from the thermodynamics of cooling—specifically whether forced convection can replace natural processes—to the durability of superconducting components and power cables in high-radiation environments. Logistically, the transport of fuel from Earth and the physical relocation of a reactor from a rocket fairing to its operational site are non-trivial. Finally, the reactor must survive the harsh interaction between regolith dust, magnetic fields, and solar activity, while ensuring that activities like venting do not irreparably damage the local vacuum.

A fuller list of the uncertainties associated with fusion on the moon is given in Chap. 8.

Chapter 7
Conclusions

Contents

7.1 The Case for Fusion Power on the Moon is Strong

The potential for fusion energy on the Moon demonstrates promise, primarily due to the lunar environment's inherent advantages. The vacuum conditions, ultra-low temperature locations, local abundance of fuel for the long term and reduced gravity. These attributes collectively contribute to creating an optimal environment for fusion reactors. These factors provide a conducive setting not only for the operational efficiency of fusion processes but also for the long-term sustainability of fusion energy production on the lunar surface.

Lunar fusion reactors are not as sensitive to fluctuating governance arrangements regarding decommissioning compared to their fission counterparts. This characteristic is crucial given the nascent and evolving domain of responsible space governance. Fusion reactors ameliorate some of the risks associated with international agreements and regulatory changes that might impact the decommissioning processes of fission reactors. Fusion reactors also eliminate the need to launch fissile material over other countries, addressing a critical risk factor in space missions involving fission reactors.

We suggest that the strongest business case for the lunar economy rests on helium-3 mining and export to Earth. ^{3}He is considered a highly valuable resource due to its potential use in aneutronic fusion power systems, but it has a strong terrestrial demand

W. J. Nuttall et al., *Fusion on the Moon*,
SpringerBriefs in Applied Sciences and Technology,
https://doi.org/10.1007/978-3-032-38217-7_7

even in advance of such fusion technology being commercialised. One such application is as an ultra-low-temperature coolant for certain types of quantum computer.[1] Between non-energy uses of ^{3}He and its use in D–^{3}He reactors lies the possible conversion of ^{3}He to tritium. At present that is economically illogical, but the import to Earth of abundant lunar ^{3}He could change all that. The Moon's surface is likely to contain substantial amounts of ^{3}He, making it an attractive target for mining operations—that themselves will require power on the order of 10 MW+. The extraction and export of ^{3}He could drive the lunar economy, providing a strong commercial incentive for developing fusion reactors on the Moon.

We see the possibility of the following rapid transitions of energy generation technology on the Moon in polar locations:

2026–2035	Solar photovoltaics on the year-round bright polar mountain tops with energy accumulation for short burst high-power capability
2030–2035	Deployment of fission microreactors is initially planned but subsequently further deployments are blocked by international agreement of the major space powers
2035–2045	Deployment of lunar D–T fusion tokamak—All fuel supplied from Earth
2045+	Deployment of lunar D–^{3}He fusion reactors on the moon fuelled with locally sourced ^{3}He and with D supplied from Earth

In the scenario, path above the period 2035–2045 is characterised by rapid and extensive industrialisation focussed on polar regions. It is that prospect that both motivates, and is driven by, fusion on the Moon. Wider issues of human settlement, including equatorial human tourism and settlement, are greatly enhanced in the period 2060–2070 if earlier rapid industrialisation at the poles has occurred. Whether that industrialisation will have occurred in a spirit of collaboration, open competition between free market actors or in a context of geopolitical rivalry and separation remains to be seen.

In Chap. 1 we pointed to the notion that Fusion energy represents perhaps the final step in mankind's energy journey. In recent decades the commercialization of fusion has been seen in terms of terrestrial electricity generation, but that is an arena with many competing low environmental impact options. This book gives voice to ideas around the nexus between space innovation and fusion innovation. We point to three motivations for such synergies:

1. The timescales of fusion and space exploration are well aligned. Both are moving fast and they can drive one another forward.
2. Arguably the unique selling point for fusion energy is energy density, in both the sense of most energy per kg of fuel and in terms of most energy per tonne of device. These attributes are key to the attractiveness of fusion to space (including lunar) applications.

[1] Ultra-low-temperatures reduce the risk of quantum decoherence in atomic quantum computers. Quantum computers are capable or solving mathematical (e.g. cryptographic) problems that are effectively impossible for conventional digital computers.

3. The synergy between ^3He fusion and a key lunar fuel source—noting a pathway to both that runs through non-fusion ^3He demand (in IT and medicine) and ^3He as a source of tritium for first generation terrestrial fusion power plants.

A convergence between fusion and space as two previously separate fields of endeavour is, we suggest, special and it could give rise to a future in the later 21st Century where we see Fusion on the Moon.

Chapter 8
Recommendations for Further Research

Contents

8.1 Research Programme 1—Addressing Information Uncertainties

This programme should focus on: gathering comprehensive data on the lunar environment; analysing the specific challenges and opportunities it presents for fusion power on the Moon; and understanding the nuances of the lunar surface, including its regolith dust/vacuum conditions, ultra-low temperature regions and ^{3}He fuel availability. This will be crucial for designing and optimising successive generations of fusion reactors.

8.2 Research Programme 2—Proving out the Business Case

This programme should aim to validate the commercial viability of lunar fusion reactors. It should involve detailed economic analyses, market assessments and feasibility studies to demonstrate the potential profitability of fusion energy production on the Moon. Additionally, it should explore the terrestrial supply and demand implications of helium-3 lunar mining and export, assessing its impact on the lunar economy and global energy markets at various timescales.

W. J. Nuttall et al., *Fusion on the Moon*,
SpringerBriefs in Applied Sciences and Technology,
https://doi.org/10.1007/978-3-032-38217-7_8

8.3 Research Programme 3—Technical Requirements of a Lunar Fusion Reactor

This programme should focus on the technical aspects of developing and deploying fusion reactors on the lunar surface. It should include research on materials science, engineering design and logistical planning to ensure the reactors' structural integrity, radiation resistance and operational efficiency. It should also explore innovative solutions for power transmission, venting mechanisms and operational practices to address the unique challenges posed by the lunar environment. It will be important to bring together engineers from both the fusion energy and aerospace sectors.

8.4 Specific Questions for Further Investigation

The proposition outlined in this book raises some key questions; the answers to some may already exist in the literature, while some may require substantial technical research to answer. We split the uncertainties into various categories. While we discussed the thematic areas of uncertainty in Chap. 6, we list here the comprehensive list of questions that need answering before embarking on fusion power generation on the Moon.

Information Uncertainties

1. What is the local abundance of deuterium?
2. What is the local abundance of helium-3?
3. What is the local abundance of lithium?
4. What is the local abundance of water-ice?
5. What are the mechanical properties of the local regolith?
6. What is the variance between satellite-observed and ground truth?
7. What are the thermal properties of the local regolith?

Governance Uncertainties

8. What are the legal requirements for decommissioning lunar activities?
9. What are the legal requirements concerning radioactive waste?
10. What are the legal or normative requirements for accessing lunar resources?
11. What is the current status of the Moon Treaty at time of launch?

Economic Uncertainties

12. What is the potential demand for lunar-sourced helium-3?
13. What is the potential demand for lunar-based fusion energy?

14. What is the timeline for commercial deployment of deuterium helium-3 fusion reactors?
15. Can helium-3 export from the Moon be economically justified?
16. What are potential environmental impacts from lunar activities?
17. How might the environmental challenges impact the lunar commercial activities?
18. Can the lunar economy evolve into a self-sustaining, independent commercial ecosystem?
19. Is a 'lunar space port' credible, as opposed to an orbital port?
20. Is the assumption that space tourism will not drive fusion on the Moon valid?
21. What are the anticipated terrestrial prices for ^{3}He in the period 2040–2070?

Operational Uncertainties

22. What buffer stores of reactor-consumables are required?
23. What are the common modes of failure?
24. What is the level of maintenance automation of the fusion reactor?
25. What are the STEEPLE requirements regarding reactor siting?
26. What other activities are being conducted on the Moon at the time?
27. Is the posited need for MW scale energy sources in PSR locations correct?
28. What are the risks and issues of damaging heat transfer to ultra-cold bedrock?

Design Uncertainties

29. Does the reactor require a breeder blanket?
30. How is reactor pulse energy stored?
31. What is the launch system?

Technical Uncertainties

32. Is fusion fuel transport from Earth trivial in terms of propellant?
33. What is the cooling load required?
34. What is the cooling/heat transfer regime, conductive, radiative or mixed?
35. What is the critical temperature of superconducting components?
36. What is the state of fusion reactor design at time of design?
37. Can forced convection replace natural convection processes?
38. How is the reactor transported from rocket fairing to operational site?
39. What are the reactor input requirements?
40. What are the reactor pulse properties expected to be?
41. What impact, if any, will regolith accumulation have on reactor operation?
42. What impact, if any, will solar activity have on a reactor sited in a PSR?
43. What risks of manufacturing/transporting a reactor at STPs for use in a PSR?
44. How long, flexible and radiation resistant are power cables in the lunar environment?

45. Does venting have any deleterious effect on local vacuum, and under what conditions?
46. How will the regolith interact with ambient and artificial magnetic fields in the work area?
47. Is wireless transmission of power viable in the lunar context?

Index

W. J. Nuttall et al., *Fusion on the Moon*,
SpringerBriefs in Applied Sciences and Technology,
https://doi.org/10.1007/978-3-032-38217-7

W

X

Kolibri 360 GmbH
Ferdinand-Jühlke-Straße 7
99095 Erfurt, Deutschland
produktsicherheit@kolibri360.de